Scanning Negatives and Slides

2nd Edition

Sascha Steinhoff

Scanning Negatives and Slides

Digitizing Your Photographic Archive

2nd Edition

rockynook

Sascha Steinhoff, www.sascha-steinhoff.de

Editor: Gerhard Rossbach
Copyeditor: Cynthia Anderson
Layout and Type: Steffen Kulpe
Cover Design: Helmut Kraus, www.exclam.de
Translation: Wilfried Bittner (1st Edition)/ Jeremy Cloot (2nd Edition)

Printer: Friesens Corporation, Altona, Canada
Printed in Canada
2nd printing, October 2009

ISBN: 978-1-933952-30-7

2nd Edition © 2009 Sascha Steinhoff
Rocky Nook, Inc.
26 West Mission Street Ste 3
Santa Barbara, CA 93101-2432
www.rockynook.com

First published under the title "Digitalisieren von Dias und Negativen"
© dpunkt.verlag GmbH, Heidelberg, Germany

Library of Congress Cataloging-in-Publication Data

Steinhoff, Sascha, 1975-[Digitalisieren von Dias und Negativen. English]
Scanning negatives and slides : digitizing your photographic archive / Sascha Steinhoff. -- 2nd ed. p. cm.
Includes bibliographical references and index.
ISBN 978-1-933952-30-7 (alk. paper)
1. Photography--Digital techniques.
2. Scanning systems.
3. Slides (Photography)
4. Photography--Negatives.

I. Title. TR267.S7413 2009 778--dc22

2008054557

Distributed by O'Reilly Media
1005 Gravenstein Highway North
Sebastopol, CA 95472-2432

Contents

Preface

Dear Reader,

Digitizing slides and negatives is still a hot topic, which is why this book is already in its third edition in Europe and its second in the US. Just about every home has its shoeboxes and slide magazines (and occasionally more professional repositories) full of analog photographic treasures, just waiting to be rediscovered. The advantages of digital image processing are widely accepted these days, and anyone who owns a scanner can use it to digitize their slides and negatives. This book will illustrate in detail all established scanning and image processing techniques, and will help you to quickly achieve great results. Your choice of scan software is just as important as good quality hardware, although the choice of commercial programs currently on the market is limited to VueScan or SilverFast. In addition to these two specialized programs, most scanner manufacturers bundle a proprietary software package with their scanner products, although these are not usually particularly powerful or easy to use. Nikon Scan is the exception to the rule.

Just as in the digital photography world, the trend in scanning is moving towards the use of RAW data formats. RAW files include the complete, unaltered image data captured by the scanner's image sensor (at least in theory), but scanning is a complex subject and has plenty of its own idiosyncracies. Please take the time to view the DVD included with this book, as it includes many high-resolution sample scans created using film and flatbed scanners from all the major manufacturers, and clearly illustrates the differences in scan quality that can be produced. I would like to thank all the readers who have provided me with feedback so far. You can contact me any time through the www.slidescan.info website, and I am looking forward to reading all your new comments and suggestions.

Sascha Steinhoff   Bangkok, January 2009

Picture: Siegfried Gromotka, um 1960

6 × 6 original negative

Enlarged crop

Darkrooms are passé: the rise of digital image processing has made fumbling around with chemical baths a thing of the past. Modern technology helps to make even 50-year-old negatives usable.

Manual correction

RAW scan

The rather flat RAW scan is just the first step. The final image is developed by stretching the levels curve and by correcting the tonal curves.

Introduction

For some time now, the digital revolution has included photography; the presence of conventional cameras in photo shops is dwindling. Unjustly so, because photographers who want to continue to work with film can also enjoy all the advantages of digital image processing by using a high-quality scanner. Owners of an archive of slides or negatives can also benefit from digitizing their existing collection. Despite the omnipresence of digital photography, there are still good reasons to shoot on film. This chapter will compare the various working methods.

1

Contents

Images: Nikon, Fuji

▲ For negative film, the processing lab is the key factor that determines the result of the printed images.

1.1 Analog and Digital Workflows

Analog Workflow for Negative Film

Over years, or even decades, many photographers accumulate boxes filled with countless negatives (although more professional storage containers are available, too). The classic workflow for negative film is as follows:

▸ Exposure – although this is not as critical as with slide film
▸ Developing and printing
▸ Proper archiving

However, there are some drawbacks to working with negative film:

▸ Negatives age and fade over time
▸ Negatives scratch easily
▸ Each trip to the lab causes new scratches, and wear and tear on the negatives
▸ It can be difficult to locate a particular image within a large archive
▸ Customized image processing is possible only in one's own darkroom

The lab work is crucial to the final result. But many photographers neither do their own processing nor do they deal with professional labs. Therefore, it is questionable whether the printed image meets their expectations. Even if the lab does a good job, there is still some latitude for variations.

The default settings of one-hour labs are generally suitable for typical snapshots from a holiday on the beach. It is a different story for photos taken in nonstandard lighting conditions. High-key, low-key, or sunset pictures often do not work well with the lab's default settings. Standard lab settings are adjusted for an average gray value, which can ruin the intended mood and render the prints unusable. Even when reordering a print from the same photo lab, the result is rarely identical to an earlier print. Only in a custom darkroom can all the parameters be controlled

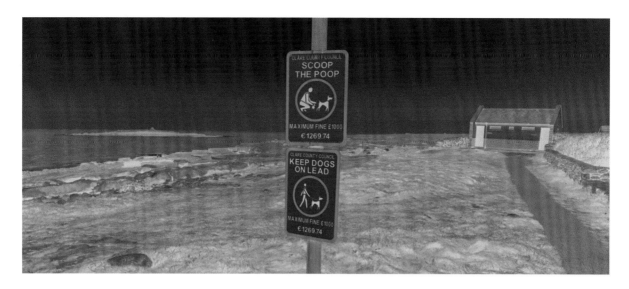

to create the desired image. To make a long story short, only with a personal darkroom can one benefit from all advantages of negative film.

Analog Workflow for Slide Film

The typical workflow for slide film differs from negative film in a few important points. It looks roughly like this:
‣ Accurate exposure is necessary; slide film is not forgiving
‣ Filmstrips are cut and framed
‣ The framed slides are filled into slide trays
‣ The images can now be projected or viewed on a light tablet

This conventional practice has disadvantages:
‣ Slides age and fade
‣ Slides get scratches, although they are less delicate than negatives
‣ For frequent slide shows, duplicates are needed because slides fade easily in the intense light of the projector; this is especially a concern for Kodachrome slides
‣ It is time-consuming to find a particular slide in a large collection
‣ Rearranging a slide show takes a lot of effort; to run more than one slide show simultaneously, duplicates are needed
‣ It is almost impossible to post-process slides

The biggest problem with slides is that they can hardly be post-processed. Poor exposure or a color cast can only be corrected with great difficulty. On the other hand, an underappreciated advantage is that the lab can do very little wrong during the processing of slides. The result is pretty close to what the camera produced. This tight control over the created image is a significant reason why ambitious photographers prefer slide film to negative film.

▲ The untreated negative shows inverted colors: on film, the white signboard is black.

▲ After inverting the colors, the signboard becomes white, but there is still the orange masking.

▲ When converting to positive, the color characteristics of the films need to be considered.

Hybrid Workflow: Shooting Analog, Scanning, Digital Processing

In recent years computer technology has advanced dramatically. Today, a standard PC actually delivers what the computer manufacturers promised more than 10 years ago: high-quality digital image processing. For those who still shoot on film, the question remains how to best get the slides and negatives into the computer.

A standard PC and a film scanner are all the hardware needed to achieve this, and the investment is manageable. The key to good quality scans is in the skill of the user. Satisfying results can only be achieved with extensive study of the matter.

Of course a quick scan is possible, but the results generally will be of poor quality. Photographers who spend a lot of effort to take quality pictures with film cameras will be unable to avoid familiarizing themselves with the subject of scanning. Only then will they be able to transfer the quality of their film-based images to the digital world.

Just like the skill of taking pictures, the skill of scanning film needs to be acquired first, in order to get good results. Sufficient time for acquiring this skill should be allowed.

This mixed workflow – shooting analog, scanning, digital processing – is also called hybrid photography. It offers all the advantages of digital image processing:

▸ "Digital negatives" don't age, scratch, or fade
▸ A well-maintained image database can retrieve images in a matter of seconds
▸ Digital images are easy to edit
▸ Digital images are easily turned into slide shows or galleries
▸ Digital images are easily sent over the Internet
▸ Digital images don't need to be touched up by the photo lab; the photographer has almost full control over the final image
▸ It is possible to make lossless backup copies of the images

▼ The film processing lab has no margin for variation: with slide film, the result from the lab is exactly what the camera produced.

Images: Nikon, Fuji

However, there are a few disadvantages:

▸ Scanning is time-consuming and fairly complex
▸ The image information of the film cannot be read without loss; unavoidably, there are losses in the transfer from analog to digital
▸ Scratches, dust, and other blemishes cannot always be removed from the film automatically during the scan (post-processing is a solution here)
▸ Not all types of film are equally suitable for scanning

To sum up, there are many good reasons for digitizing your film material, so don't put it off much longer. The latest high-grade film scanners do a fine job of extracting the image information from the film material. Unlike with an analog workflow, all relevant parameters for image quality can be defined on the computer. When ordering prints from a lab, using digital images leads to significantly better results than working from negatives. In addition, there are no more annoying scratches and fingerprints, which usually further damages the negative with every trip to the lab.

In the early days of digital photography, when digicams were as expensive as picture quality was poor, there were many photo enthusiasts who propagated hybrid photography as the best solution for quality-conscious people. Since then, however, much has happened in the world of photography and there are only a few who still shoot pictures on old-fashioned film. We are living in a fast world, and nobody has the time to scan filmrolls all evening. Due to this, everybody has a digital camera and hybrid photography is only important for archiving purposes.

▼ Home users can explore digital image processing with a desktop film scanner.

Images: Nikon, Fuji, HP

Digital Workflow: Shooting and Editing

Purely digital workflow requires that images be generated and processed digitally. Technically this is desirable because it eliminates the potentially degrading analog to digital conversion between image and image file. However, for practical photography, there are a few limitations:

▸ There is still no common standard for a "digital negative"
▸ Existing equipment for analog SLR camera systems can only be used with certain restrictions
▸ High-resolution digital cameras with a full format sensor are significantly more expensive than analog cameras, although prices keep dropping (in addition, there are fewer analog cameras available)

Still, the advantages are undeniable and have led to the unstoppable success of digital photography:

▸ Digital images can be edited directly on the computer
▸ Subsequent quality-degrading steps of Analog/Digital (A/D) conversions of the picture are eliminated
▸ No more burning film – memory cards are much less expensive in the long run
▸ The picture can be checked immediately after it has been taken – good cameras have a built-in histogram display for checking the exposure

▲ Digital cameras no longer capture the image on film, but through either a CMOS or CCD sensor chip (here Nikon D2H). Sooner or later, this technology will completely replace 35mm film.

An often-underappreciated benefit of the hybrid workflow over the digital workflow is having a tangible piece of film. In addition to the digital image file, there is always the analog backup. As long as there is no cross-platform standard for file formats and storage media, this is a major advantage over the purely digital workflow. With technology constantly progressing, digital images must be periodically transferred onto different backup media. Neglecting this step can cause the data to become unreadable or completely lost. As an example, in the early nineties home computers had floppy disks, then came CD-ROMs, and now we have DVDs, external hard disks, and USB (Universal Serial Bus) thumb drives. A current computer system can no longer read the types of media that were common in the mid-eighties.

Analog images must merely be stored dry and free of dust; no other measures are needed. However, if you store both a slide tray and a DVD with image files in the attic and 30 years later someone finds them, the chances are low that the image files on the DVD can still be used — provided that the DVD can still be read at all. The slides, on the other hand, will be slightly faded, but with recognizable images.

The digital photography boom has raised the level of image quality and availability to a level unimaginable to most amateurs in the old, analog photography world. The price for this ease of use is often paid with flawed or faulty data storage, and there is still no digital storage method that is as robust as conventional camera film. In 2008, for example, lost pictures taken over 70 years ago by the legendary Magnum photographer Robert Capa turned up largely unharmed after decades

▲ From the digital camera or scanner directly to the lab: now most photo labs offer a digital printing service.

of storage in a suitcase. Digital images would certainly not have been able to survive that long without being professionally archived. There are services that offer to expose your digital pictures on slide film for archiving purposes. But to be completely on the safe side, neither analog nor digital photography is useful anyway. For maximum durability, you should do as in Egypt: carve your information in a solid piece of stone and bury this in a protected place in your garden.

▲ A digital workflow allows lossless transfer of images from camera to PC.

▼ The digitizing of archive material is of ever-increasing importance (Image: Siegfried Gromotka, around 1960).

Image: Nikon

▲ Nikon D300: 12.3 megapixels
and continuous shooting of six
frames per second.

Image: Nikon

▲ The quality of the lens essentially
determines the effective resolution, both for
film and digital cameras.

1.2 Alternatives to the Film Scanner

Scanning filmstrips with a film scanner is time-consuming and has a steep learning curve. Because not everybody has the time, knowledge, or hardware needed, let's discuss the possible alternatives.

Megapixel in Comparison:
DSLR, Film Scanner, 35 mm Film

Quite often, only the nominal resolution is considered when comparing DSLR, film scanner, and 35mm film. According to this criterion alone, analog cameras would come out on top. However, it is not that simple. For a DSLR, the sensor and the lens determine the resolution of the final image. This limiting factor is always the weakest link in the camera-lens combination: if you mount an inferior lens on an excellent DSLR, the sensor cannot utilize its maximum resolution.

It's all about the lens

Only high-grade lenses have enough quality reserves for a good DSLR. The resolution of current 35mm DSLRs is between 6 and 16 MP. Lens aberrations are very noticeable with DSLRs. Unlike an image from an analog camera, the digital image is usually examined at maximum size on the computer monitor. To notice such aberrations on a 35mm slide, it would take a high-power loupe. Scanning film is a more complex issue. Here the image is generated in two steps: first, the subject is shot on 35mm film, and then this image is scanned. Therefore, we have to look not only at the quality of the material to be scanned, but also at whether the scanner is capable of capturing the image adequately. For the resolution of the film – let's say 35mm film – the combination of film and lens is crucial. A high-quality, fine-grain 35mm film has a resolution corresponding to 40–60 MP. Standard film has a substantially lower resolution, which corresponds to around 20–30 MP; but even that resolution can be achieved only with high quality lenses. Simple lenses achieve a resolution of approximately 10 MP; only high-grade lenses will surpass 20 MP. The popular amateur zoom lenses are typically inferior to the optical performance of comparable prime lenses. More than likely, the majority of amateur pictures have a resolution of less than 20 MP.

Effective Scan Resolution is Crucial

The effective resolution of a film scanner is potentially a bottleneck that can limit the resolution of the image. Currently most high-quality film scanners have a resolution of 4000 dots per inch (DPI or dpi), which corresponds to around 20 MP. If you choose a film scanner that not only promises this resolution on paper, but can also scan it optically, then loss in image resolution will not be noticed.

There are already 35mm film scanners with 7200 dpi, which corresponds to a resolution of 70 MP. To fully utilize this dpi, the slide or negative has to be of excellent quality, which means a fine-grain film exposed with a very sharp lens. The same is true for the scanner. It not only has to have a high-resolution sensor but also excellent optics.

I am not aware of any current desktop film scanner that in actual use resolves more than 4000 dpi. For most source material, a resolution of 2900 dpi would be sufficient. Also, keep in mind that file sizes get very large with increasing resolution, especially with medium and large format scans. With file sizes of over 200 megabytes (MB), image processing will be very difficult on a regular home PC. When choosing the suitable scan resolution, the entire workflow has to be considered.

Meanwhile, professional DSLRs have approached the nominal resolution of far less expensive analog cameras. For example, the Canon EOS 1Ds Mark II is fitted with a full-frame sensor which has a nominal resolution of 16.7 MP. According to the tests in www.dpreview.com, the actual resolution is only 2800 lines horizontally and 2400 lines vertically. Thus, the high nominal resolution translates into an actual, measured resolution of only 6.7 MP. A similar ratio between nominal and measured resolution applies to all models of digital cameras.

6 MP Easily Beats 20 MP

When comparing image quality between scanned analog images and digital images, it is not sufficient to only consider nominal resolutions. The degree of quality loss that occurs during film scanning depends on various factors such as scan density and film flatness. But keep in mind that there is always a loss when converting analog to digital. Due to this, even a 6 MP DSLR is superior to a 4000 dpi (20 MP) scan in most cases, despite the nominally lower resolution.

The nominal resolution of a scanner is often the single most important purchase criterion for the layman. For this reason, manufacturers (especially those of lower quality scanners) tend to advertise their products using improbably high resolutions. Plustek, for example, advertises resolutions of up to 7200 dpi which are rarely actually achieved. Furthermore, a resolution of more than 4000 dpi is not actually necessary for most practical purposes. Artificially high resolutions also lead to unnecessary use of additional storage resources.

▶ Looking just at the numbers for the near future, film scanners will nominally remain far superior to amateur DSLRs – such as the Nikon D70 or D70s.

6.1 Megapixel
Nikon D70

2900 spi = 11.3 Megapixel

4000 spi = 21.4 Megapixel

5400 spi = 39.1 Megapixel

Digital Single Lens Reflex Cameras

Some years ago there was an ongoing battle among photographers as to whether analog or digital cameras would be the best choice for superior quality. According to current sales figures, this battle has been won by digital cameras, but the technology still has some limitations.

Digital cameras are without discussion the best for product shots and other technical pictures. For portraits, it is a complete different story. On the one hand, the skin tones of the digital image are rarely as expected; on the other hand, the faithful and accurate digital reproduction is not always flattering to the subject, since even the slightest skin imperfection will be precisely recorded. Therefore, in the area of portrait photography, many professionals still prefer film for its very particular, slightly grainy appearance.

The other weak point of digital photography is the always touchy subject of data backup. In the hybrid workflow, there is always the original filmstrip to go back to, but with digital photography all will be lost should the computer crash one day. It remains to be seen how "future-proof" digital images will be.

A 50-year-old negative can still be processed today without any problem. This may in the future not be the case for all of today's popular file formats. Image files without backup can be quickly and permanently lost in a computer crash, while slides and negatives have shown more robustness over the last decade.

And last but not least, there is one crucial reason why even the latest digital camera will never be able to replace a scanner. It has no go-back button; you cannot use it for pictures of your own childhood or anything else that happened in the past. Due to this, you will sooner or later need to make use of a scanner. Digital cameras are great, but not of much use when you are digitizing your analog archives.

▲ The sensor of a DSLR does not reflect light the same way as film (here Nikon D2H). For this reason, the TTL metering of older flash units does not work on modern digital cameras.(Image: Nikon)

Images: Epson

◀ The Epson Perfection V700 is not quite as good as a film scanner, but it is clearly better than a traditional flatbed scanner. For medium format, it is a viable alternative.

Flatbed Scanners with Transparency Adapters

Flatbed scanners have developed enormously over the last few years. They were originally developed for scanning documents. But, flatbed scanners have achieved a resolution which is unnecessary for scanning documents alone. There are many models with transparency units that allow the scanning of film. At present, this technology remains inferior to specialized film scanners.

However, flatbed scanners are faster than film scanners, an advantage not to be underestimated. Though flatbed scanners advertise resolutions of up to 4800 dpi, tests show that this is incorrect. Flatbed scanners that boast a nominal resolution of 4800 dpi can actually only capture 1700 dpi from a transparency. At that resolution, a 35mm slide can only be enlarged up to 5" × 7" at 300 dpi. Also, the color and contrast do not come close to what you can expect from a good quality film scanner. Please compare the scanned files on the DVD accompanying this book.

For scanning medium- and large-format film, the maximum resolution is not so critical, since the film size is several times larger than 35mm film. Average flatbed scanners have a glass stage for holding the originals. However, this glass plate reduces the optical performance of the scanner. In a dedicated film scanner, there is no glass between sensor and medium. This is the reason that, for the foreseeable future, conventional desktop flatbed scanners will not match the performance of dedicated film scanners. Epson tried to improve the unsatisfactory performance of flatbed scanners with a new design approach. The Perfection V700 has a separate lens for scanning film. It achieves an effective resolution of 1920 × 1770 dpi and a maximum density (Dmax) of 3.4, which is just below a good film scanner.

Even the best flatbed scanners cannot achieve the quality you can obtain using a specially designed film scanner, but they are more flexible when it comes to the sizes and formats of originals that can be scanned. If you only want to scan negatives and slides, you are better off using a film scanner, whereas flatbed is better if you need to scan varying formats. Ideally, you should have both types for use in different situations.

Nikon Coolscan 5000: The dedicated filmscanner points out details very clearly and produces consistently vivid colors.

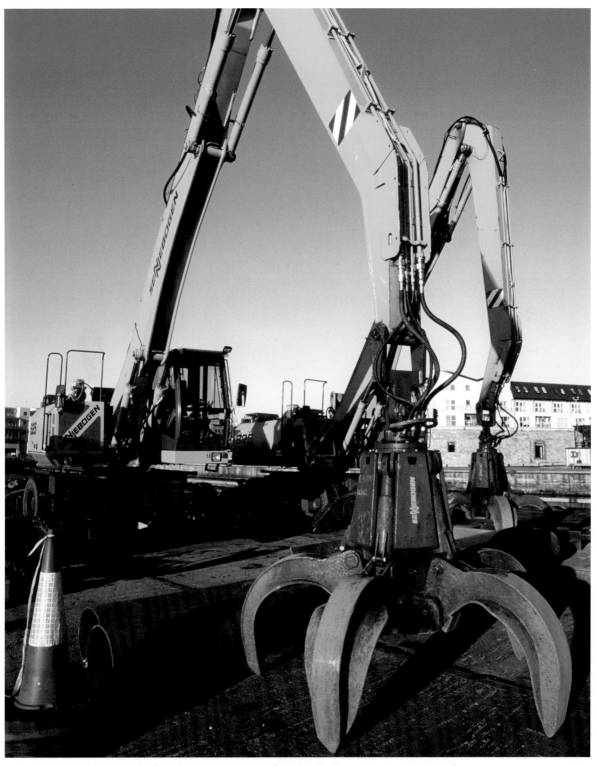

Epson V750 Pro: At first glance, this flatbed scan appears to be of good quality, but a direct comparison to the Nikon (left side) shows a lack of sharpness and less accurate color reproduction.

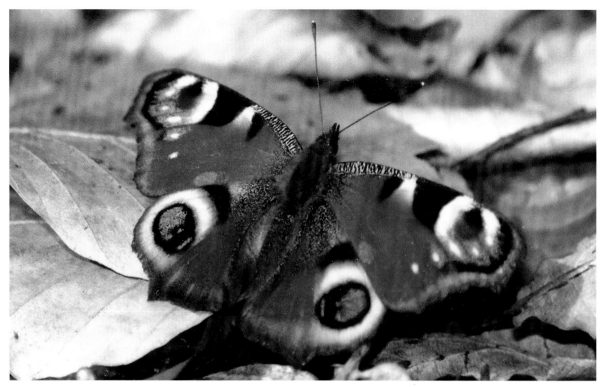

Scan of a 4" × 6"-print: Leaves barely show details.

Scan of the corresponding negative: Even the subtle grain of the leaves is visible.

Scanning Prints

Occasionally, there are discussions on various forums about first making a print from a negative that is as large as possible, then scanning the print with a flatbed scanner. This method is one possibility – but the quality is noticeably worse than when scanning directly from the negative. Scanning the print is not inexpensive, either. Technically speaking, it is an analog-analog-digital conversion. The film is turned into a print, and then the print is scanned. With each step, some information gets lost, which is seen in the final image file. The additional step of printing causes noticeable quality loss. It is equivalent to playing a song on a record player and then recording it with a microphone from the speakers. Although technically possible, it is not optimal in terms of quality.

The quality may be acceptable if the image size is reduced by scanning, but this method cannot match the quality of a scan from the original slide or negative. Scanning the original film – if it is still available – is definitely the better solution. If the original slide or negative is still available, it should be scanned directly. Even though it can yield enlargements up to poster size, enlarging scanned prints should be avoided. Reductions in size usually yield better results than enlargement; even a 1:1 scan of a print is not ideal.

As can be seen from the following images, fine details get lost when scanning from a print. This is clearly evident in the texture of the leaves. Compared to a scan from a negative, a scan from a print suffers a noticeable loss of information. The print was scanned at 1200 dpi; the negative, at 2900 dpi. Again, the print should be scanned only if the slide or negative is not available.

You should always keep your original negatives or slides, as they are the only source of the complete original image information, regardless of how good your scan may be. Experience has also shown that analog images often have a longer shelf-life than digitally stored images.

Image: Novoflex

◀ Novoflex slide duplicator: suitable for 35mm and medium format.

Shooting a Projected Slide with a Digital Camera

Another option is to project a slide onto a wall and photograph it using a digital camera. The image will then be digitized. The concept is a good one, though not new; it was already being discussed in the old analog days. The idea is to project two slides on top of each other to create a sandwich slide. This way, one would not have to invest in a dedicated slide duplication attachment. Today, as in the past, the results are rather poor. The images are flat, dark, and soft. The slide loses a large degree of its sharpness when projected onto a wall. Therefore, for good quality digitized slides, this is not a suitable method.

Duplicating Slides with DSLR and Slide Copy Attachment

A good DSLR with a macro lens and a slide duplicator can indeed be used to digitize slides. A bellows can also be added. However, the quality of the images resulting from this combination does not match that of using a film scanner. In addition, there is no hardware-based scratch and dust removal available. For now, this is not a real alternative to a film scanner. If you need to digitize slides only occasionally and already have suitable equipment, it is worth a try, especially since a slide copy attachment is relatively inexpensive. Although this technique is still in its infancy, it does have potential. It is blazingly fast. No film scanner comes even close.

Film Processing with Photo CDs

The most convenient way to generate files from film is through a Photo CD, because the lab does all the work for you. Photo CDs are offered inexpensively with film processing. Even if you do your own scanning, it can make sense to add a Photo CD to your developing and processing order. For a small charge, you get an overview of the images and can select those images that are worth rescanning at a higher resolution. It saves time, and the image quality is sufficient to let friends and relatives have quick prints made from the images.

Kodak Picture CDs

Image: Kodak

▲ Inexpensive at an adequate quality:
Kodak Picture CDture CD

A picture CD from Kodak is very inexpensive, but only provides images with a resolution of 1536 x 1024. An entire roll of film adds up to around 25 MB, so a disk can hold 26 rolls. Even though the quality is only good enough for 3R to 4R prints, for this small extra cost you can see whether any images are worth scanning yourself at a higher resolution. Picture CDs are typically offered only when developing negative film. The cost is around $2.50 per roll and about three times that for slides. For more demanding needs, the Photo CD is a good option. It is more expensive than Picture CD and is available for slides and negatives. The image format used is called Image Pac. This format is supported by most image editing software, either directly or through plug-ins. The images are stored at five different resolutions simultaneously, ranging from 128 × 192 to 2048 × 3072 pixels. A Photo CD costs around $5.

Even third-party service providers can scan images onto Kodak's Photo CD. They typically charge per scan. The Pro-Photo CD, which costs around $11 per CD, is geared more toward professional use. With its high resolution of 4096 × 6144, it nearly matches the quality of the original. The lab devotes a greater processing effort compared to the Photo CD. However, currently, only around 25 images fit on one Pro-Photo CD.

Kodak Photo CDs are an established standard and can be displayed on most DVD players, which is handy for displaying images on TV without much effort. You can find more information about Kodak Photo CDs at www.kodak.com and www.tedfelix.com/photocd. However, with the continuing spread of digital cameras, it seems inevitable that these Kodak formats will be replaced before long.

Fujicolor CD

Fuji can produce scans from 35mm or Advanced Photo System (APS) negatives in JPEG format. The resolution is 1024 × 1527 pixels with a color depth of 24 bits. The cost per Fujicolor CD is around $2.50. It is a counterpart to the Kodak Picture CD and therefore is also not for the quality-minded.

Comparison: Photo CDs vs. Manual Scans

With all current formats of photo CDs, photos undergo extensive processing. Although you are spared the trouble of processing yourself, you give up control over critical parameters of image capture. Since the possibility of post-processing JPEG images is limited, you have to accept the results you get if you do not want to rescan the images yourself.

For shots in normal lighting, the results from the photo CDs are generally usable; but scanning yourself can yield much more from the originals. The images on the following pages clearly show how much image information gets lost in a standard scan from a chain lab. In the example shown, the lighting wasn't even difficult.

The scan from the lab is of only moderate quality because the tonal range is clipped. This is especially noticeable in the foliage in the background, which disappears into a featureless blackness. This cannot be rectified, since the compressed JPEG image from the CD doesn't respond very well to image processing. Also, the possibilities for enlarged crops are limited due to the small resolution (1024 × 1536 pixels).

Only a manual scan will reveal all the unused image information in the original, but it takes a correspondingly long amount of time. Using standard corrections such as setting the black/white/gray points and the analog gain before the scan, and following that with the shadow/highlight filter in Photoshop, can give a much better-looking result.

Whichever scanning service or lab you decide to use, the quality of the results will depend almost entirely on the quality of the work done by the individual technician. Making good quality scans is time-consuming work, and a large, mass-production lab will rarely produce scans of high enough quality to satisfy serious photographers. If you scan your own images, you can control all of the most critical parameters yourself.

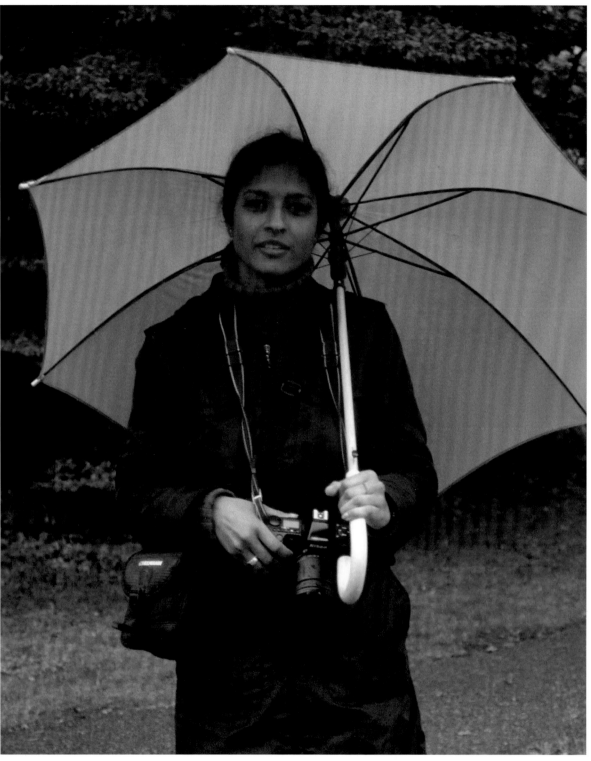

Photo CD: The fine details of the slide get lost in the scan from the drugstore. With JPEG, even post-processing can't recover anything.

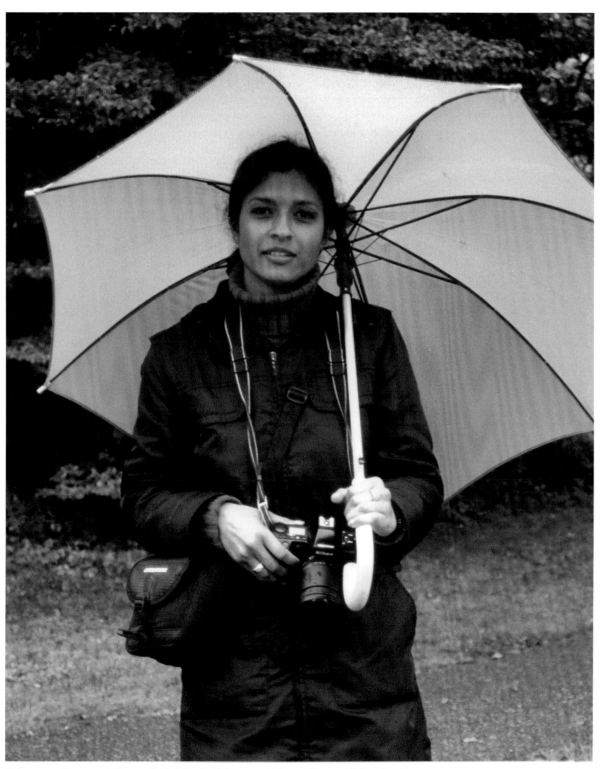

Only a manual scan can bring out the full details of the original. The tonal range is fully utilized because it isn't restricted by any automatic mode.

Professional Scanning Services

Subsequent scanning of film material for improved quality is time-consuming and bothersome. There are good arguments for using professional services, such as www.slidescanning.com, www.digmypics.com, www.imagerylab.com, and www.gemega.com. These services usually only scan processed film, since they don't have their own photo labs. Although the cost for a single image may be manageable, if you have a large number of slides or negatives to scan, total costs can add up to a substantial amount. Prices for processing vary markedly, as can the quality of the results. Unlike the CDs offered by photo labs, specialized services support many different resolutions and file formats. Almost anything is possible, but the price increases with the resolution.

The standard file formats TIFF and JPEG are common. High-resolution images in TIFF format are of high quality and are well-suited for postprocessing. RAW formats are not offered very often, since no dominant standard has emerged so far. A scanning service should automatically remove scratch and dust with Image Correction and Enhancement (ICE) or Film Automatic Retouching and Enhancement (FARE). Some companies also offer correction schemes to revive the colors of faded originals.

Scanning services are a good choice if you do not have the time or the inclination to deal with a scanner yourself. This would be the case if you want to completely digitize your old stock and intend to shoot new images digitally. To limit costs, you should sort through your archive beforehand. Many photo enthusiasts have collections of 10,000 or more slides. Only in rare cases would all images be good enough to justify digitizing an entire archive. If only a few hundred originals need scanning, buying a scanner is not cost-effective.

* Before you hand over large numbers of slides or negatives to a lab for scanning, you should do a "test run" with a few selected images, in order to make sure that scan quality is up to standard. Even if such a test produces good results, it is always prudent to divide your originals into smaller lots, thus minimizing the risk of losing your images at the lab or in the mail.

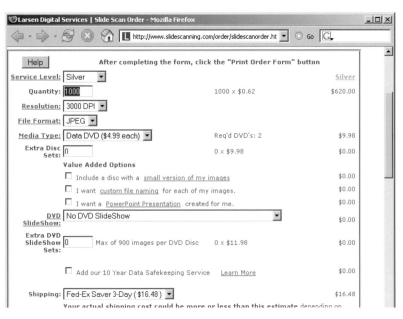

▶ The small scanning cost per single image adds up to a substantial amount for larger quantities.

Performance Characteristics of Film Scanners

This chapter discusses all the essential parameters for evaluating a film scanner. Apart from the numerically definable aspects such as resolution and dynamic range, the not so easily quantifiable quality of the optics is critical. Because specifications on paper rarely match actual performance, test reports should definitely be read before purchasing a scanner. The DVD for this book contains sample scans from almost all current film scanners. These can be used to compare scanners directly, or for an overview of the different quality levels of the various models. Detailed test reports in different languages can be found on the Internet at www.filmscanner.info.

2

Contents

▲ The original size of a 35mm slide is only 24 × 36mm, but you can print posters from it as large as 12" × 18".

2.1 Key Specifications of a Film Scanner

Nominal Resolution

The resolution of a film scanner defines its ability to record even the smallest details of the original image scanned. Usually it is defined in pixels per inch (PPI or ppi). Occasionally the units DPI (dots per inch) and SPI (samples per inch) may also be used. Hence, a film scanner with 4000 ppi can resolve 4000 pixels per inch. The higher the resolution of the film scanner, the larger the generated images will be. This is a good thing, because high-resolution images can still be reduced any time. However, the higher the resolution, the larger the image file becomes. A modern 35mm film scanner produces 16-bit TIFF files with more than 110 MB per image. If you scan a few thousand images, storage requirements will grow accordingly. Even if you do not plan to use the maximum resolution for normal scanning, it should still be as high as possible to allow for tight crops.

2000 dpi is Sufficient in Most Cases

With a print resolution of 300 dpi, a 4000 spi scan can be printed up to 12" × 18". In theory, a 4000 spi scan can be printed with a resolution of 300 dpi at up to 12" x 18". In practice, however, most originals are not of high enough quality, and even the smallest flaws can no longer be effectively touched up at such high resolutions. At the end of the day, 2000 spi is sufficient for most scans.

Interpolation is Worthless

Only the scanner's actual optical resolution is of interest. Catalogs often state the interpolated resolution, which is considerably higher. Interpolated resolutions are meaningless; the optical resolution of the scanner is only mathematically up-sampled, resulting in artificially inflated image files. Yet, they only contain information that the scanner can physically record with its optical resolution. If you want to print at a larger size, you should use an image editor for up-sampling. For that, you do not need a special scanner. It is not recommended to interpolate the scan data directly in the scanning software. It is better to scan at the scanner's native resolution, at which it delivers the best image quality.

Hands-on Tip:
With Small Formats, Determine Output Size First

If the scanner automatically scales the image, the film grain will often look more pleasing than that produced from a high-resolution scan down-sampled to the desired size in an image editor afterwards. This effect is very pronounced with Nikon scanners.

4000 spi scan – maximum crop for a full-page print at 300 dpi.

PPI, SPI, and DPI: Units of Resolution

A scanner's resolution is specified in PPI, SPI, or DPI. All these terms are units for resolution per inch. Which one to use depends on the field to which it refers.

PPI = Pixel per Inch   This unit refers to an image display on a monitor. It describes the number of pixels for every inch of the image on the screen. A pixel is the smallest unit the monitor uses to represent an image.

SPI = Samples per Inch   This unit refers to the number of samples (scanning points) per inch on the original that the scanner can record. A sample is the smallest image spot the scanner can detect.

DPI = Dots per Inch   This unit refers to the resolution of a printer. It defines the number of printed dots per inch the printer can deposit on the print medium. For an inkjet printer, one dot corresponds to its technically smallest possible ink droplet. The printer cannot deposit less than one droplet. Consequently, a high-resolution inkjet printer must be able to generate very small droplets.

Unfortunately, few people adhere to these textbook definitions. In this book, the terms are used according to common practice. It is only of academic concern whether the resolution of a scanner is quoted in ppi, dpi, or correctly in spi. Even popular image editors commonly confuse the terms. Since resolution is the most prominent feature of a scanner and is crucial to a customer making a purchasing decision, some manufacturers resort to printing unrealistic resolution specifications on their boxes. There are clear standards for measuring resolution, like ISO 16667 certification, but unfortunately certification is not mandatory for the manufacturers. Test reports from independent computer or photo magazines are better for providing information about resolution, since such reports measure advertising specifications against harsh reality.

PPI, SPI & DPI: Each of these terms is precisely defined, depending on its intended usage. These definitions tend, however, to be loosely or even wrongly applied in daily use. For this reason, you can generally use the term DPI to describe a scanner's resolution, even though it is not technically correct.

Resolution Swindle Eats Your Storage

If a manufacturer promises 3600 dpi, but in reality the scanner only resolves 1800 dpi, you not only get just half the promised resolution; adding insult to injury you also have to suffer large file sizes resulting from a 3600 dpi scan. Plenty of storage space is wasted without getting better pictures. For high-grade film scanners, the difference between advertised and actual resolution is usually small, somewhere in the range of 10 percent. Flatbed scanners with transparency units are a completely different story. With typical resolution reported at around 4800 dpi, on paper they surpass film scanners; but the current crop of flatbed scanners – at the time of this printing – only resolve 1700 dpi in transparency mode. In general, you should not trust resolution promises blindly.

USAF-1951 Test Target: Determining the Actual Resolution

You will need a dedicated USAF 1951 test target to determine the actual resolution of a scanner by yourself. You can order such targets from www.sinepatterns.com or www.appliedimage.com. They come in different sizes; to test a film scanner, you will need a 35mm target. Typically the targets are delivered as unframed transparencies. They can be used with film scanners as well as flatbed scanners with transparency adapters.

After you mount the target in a slide frame, it fits into a film scanner or flatbed scanner without problems. The target contains black bars of different sizes on a clear background. The bars are aligned horizontally and vertically because the scanner can attain different resolutions in horizontal and vertical directions.

Set the scanner to its optical resolution and save in a lossless file format, such as TIFF. Do not apply any image corrections or filters as they would distort the results. This is easier said than done, as some scanners have the unsharp filtering function implemented in the software or even in the hardware. That means you can untick the unsharp masking box in the software GUI, but actually the scanner still sharpens the picture. This effect happens especially with cheap scanning devices. The only thing you can do about it is to use proper software like SilverFast or Vuescan.

Once you have completed the scan, open the image file in a viewer and set the zoom factor to 100 percent. Then, interpret the displayed target elements by using the lookup table that comes with the target. This method, is, however, not particularly precise, as judging whether a pixel has been precisely rendered or not is subjective.

When comparing scanners, it is also important to remember that some manufacturers are playing with a marked deck. Most scan software allows you (theoretically) to switch off the unsharp mask feature, but many cheaper flatbed scanners leave unsharp mask permanently switched on in order to make the scan results more usable. Remember: even the most complex post-processing tricks cannot effectively compensate for a low-quality scan. All in all, the USAF target is an effective tool, for comparing scanners directly on your own. It is especially helpful to find out if your own scanner has been manufactured properly.

▲ Counting bars with the USAF-1951 test target from Applied Image. Although the target is not as accurate as required by the ISO standard, it works well for finding differences in the quality of scanners.

Exact Measurement of Scanner Resolution: ISO 16667 vs. ISO 14773

ISO 16667 is used to measure actual resolution. It is a very useful standard but quite seldom used. In contrast, ISO 14773 is not conclusive, since it merely determines nominal resolution.

Color Depth of Scanner

Color depth only describes how many bits are used to store a pixel and has no direct relevance to the quality of color reproduction. An 8-bit scanner can potentially reproduce colors better than a 16-bit scanner.

As will be shown later in Chapter 4 on the topic of file formats, maximum color depth is an important criterion for evaluating a scanner. Common color depths of scanners are 8-, 12-, 14-, and 16-bit. These correspond to the following grayscale values:

▸ 8-bit color depth: 256 (2^8) grayscales
▸ 12-bit color depth: 4 096 (2^{12}) grayscales
▸ 14-bit color depth: 16 384 (2^{14}) grayscales
▸ 16-bit color depth: 65 536 (2^{16}) grayscales

In Red, Green, and Blue (RGB) mode, each image point is generated from three values (one each for red, green, and blue). The number of bits refers to each of the three color channels. These correspond to the following number of possible colors in RGB mode:

▸ 8-bit RGB = $2^8 \times 2^8 \times 2^8$ = 2^{24} = 16.777.216 colors
▸ 12-bit RGB = $2^{12} \times 2^{12} \times 2^{12}$ = 2^{36} = 68.719.476.736 colors
▸ 14-bit RGB = $2^{14} \times 2^{14} \times 2^{14}$ = 2^{42} = 4.398.046.511.104 colors
▸ 16-bit RGB = $2^{16} \times 2^{16} \times 2^{16}$ = 2^{48} = 281.474.976.710.656 colors

Obviously, the numbers of colors for RGB are somewhat unwieldy, which is why bit-depth is more commonly used for expressing color depth. It is important to keep in mind that a 16-bit file only uses twice as much storage space as an 8-bit file but offers an exponentially higher color depth. Storing the files in RAW format makes the best use of your hard disk space. Storing the images in TIFF can take up more space, except for 8-bit and 16-bit scans, which are directly supported by TIFF. The only lossless way to save 12-bit and 14-bit scan data in TIFF is in 16-bit color space, where either 4/16 or 2/16 of the storage space will be wasted.

Color Depth in Image Processing: 8-Bit or 16-Bit?

Many image editing programs still work only in 8-bit, but a few also work in 16-bit color depth. A 16-bit color depth is technically by far the better solution, but the question is whether or not it is worth the effort. Working in 16-bit is very demanding. Not all image editors support 16-bit throughout. Furthermore, 16-bit images occupy twice as much memory as 8-bit images, which are already quite large. You would need a much faster computer, too.

For image output, 16-bit offers no advantage. That is, the human eye cannot detect any additional image information with 16-bit. It is a different case, however, for image editing. If you modify color curves, change contrasts, or adjust brightness, a 16-bit source file can contain the decisive bit of extra information to allow for certain adjustments without visible degradation. If you are not planning to do any extensive post-processing on your scanned images, 16-bit unnecessarily wastes time and storage space.

This slide has been scanned at 16-bit color depth and output as a 16-bit TIFF file.
The only other processing was unsharp masking.

The same slide saved as an 8-bit TIFF file.
There is no noticeable difference in the prints.

Density Range

Apart from resolution and color depth, density range is an important criterion for selecting a scanner. Density range describes the difference between the area with the maximum density (Dmax) and the area with the minimum density (Dmin). It is the difference between the darkest and the brightest image points that the scanner can differentiate from pure black or white, respectively. The higher the density of the original, the more brilliant and contrasty the image will be. If the film scanner has a lower density range than the original slide, the scan will suffer some quality loss. Not all nuances in color and grayscale can be captured entirely. The scanner cannot record all details in the highlight and shadow areas simultaneously. Either the shadow details will drown in darkness, or highlight details will be bleached away. Especially for slides, a high Dmax on the scanner is essential for a good image quality.

Low vs. High Density: Spot the Difference

Of course it is unfair to compare a current and high-priced scanner like the Nikon Coolscan 5000 to the Minolta Scan Dual II, which is out of production and was in a lower price range. However the sample image with the bird demonstrates the obvious disparity in quality between different scanners. In both cases, the density range published in the specification sheets is to be taken with a grain of salt. Regardless, the Nikon Coolscan 5000 undoubtedly has the higher tonal range, which shows in the image result. While the Minolta scan appears flat, Nikon brings out the brilliance of the slide nicely.

Density Range = Tonal Range
The density range is also measured as tonal range with the unit D (for Density). It can be clearly determined with the ISO 21550 standard.

In both cases the manufacturer's software was used for scanning. Comparing different scanners is like comparing apples and oranges, because the result depends not only on the hardware but also greatly on the software used. The brilliance of a slide can translate into a good scan only if the scanner has a high actual density range. Fallacious numbers from the specification sheet will be quickly exposed.

Multi Exposure: Enhance Density by Software

Despite the fact that a high density range always needs good scanning hardware behind it, it is possible to expand the scanner limitations by software. Some time ago software manufacturer SilverFast introduced its Multi Exposure feature, which basically means that the scanner takes two samples of the slide while scanning with different exposure durations. Both samples are combined by the software and you get a picture with an extended density range. This reduces grain, too. But of course your scanning time will reduplicate.

Do not mix up this technique with Multi Sampling. In Multi Sampling you process several passes, too, but all with the same exposure time. VueScan has a similar feature which is called Multi Exposure. Nikon Scan has no comparable functionality to offer.

Scan of a color slide with Minolta Scan Dual II (3.2 D)

Comparison with Nikon Coolscan 5000 ED (4.8 D)

Scanning Speed

Even using a fast scanner, scanning a complete roll of 35mm film usually takes about two hours, excluding any additional post-processing.

Scanning speed is a limiting factor when digitizing film. Generally, it takes longer to scan negative film than slides because the scanner needs to invert the negative, which requires additional computing time. For a complete scan of a single negative frame, even the fastest desktop film scanners need around one minute per frame with scratch removal enabled. Unfortunately, substantially longer times are the norm. If the scanner supports bulk scanning, this is still acceptable, since the scanner can work unattended. However, if you do many single scans, a fast scanner is a must. You can find extensive test reports for most of the film scanners on the market at www.filmscanner.info, including detailed information about scanner processing speed, provided in German, English, French, Spanish and Italian.

Most film scanners are not set up to let you quickly digitize a few pictures. Scanning takes quite a bit of time and can test your patience. Resolution and image quality of modern film scanners are already very good, but there is still plenty of room for improvement as far as processing speed is concerned. If you need a fast system now, you will have to dig deep into your pockets for the scanner and a computer powerful enough to adequately support the scanner. Currently, the fastest scanner is the Nikon Coolscan 5000 ED. You should use at least a PC in the 3GHz class with enough RAM to run the scanner at its optimal speed.

Implemented Image Corrections

Modern film scanners employ special correction methods for removing dust and scratches, and for equalizing film grain. These technologies are not luxuries but rather basic requirements for turning good originals into high-quality image files. In addition, it is possible to boost colors of faded originals and to correct shadows and highlights at the scanning stage. The best methods for dust and scratch removal involve inspecting the film surface with an infrared beam. This method cannot be replicated in software with equally good results. To make use of this feature, your scanner will need a separate infrared channel. Most modern scanners have this feature nowadays.

Apart from this, all other image corrections can be performed afterwards with an image editor. Color restoration and grain reduction can be done in Nikon Scan, SilverFast, VueScan, or in a Photoshop plug-in (which you have to buy separately). The general advantage of image corrections in the scanner and scanning software is that they happen during scanning, and therefore require less time-consuming post-processing in the image editor. In most cases, the corrections will be firmly embedded in the image file, even when using RAW. You can save time by making sure the scanner is set up properly. If you want to change the level of correction afterwards, you must scan again, which can be a pain. To speed up your workflow, it should be your goal to scan your slides and negatives once, and to do everything else in post-processing.

Supported Film Formats and Bulk Scanning

You should already know before you purchase a scanner whether you want to scan 35mm, APS, medium format, or large format film. Currently, there is no affordable scanner for processing all formats in good quality. All common desktop scanners can handle 35mm negatives and slides, but support for additional formats varies.

For example, Nikon's Super Coolscan 9000 ED scans 35mm and medium format from 4.5 × 6cm to 6 × 9cm. There are even adapters for such exotic formats as 16mm film, 24 × 65mm panoramas, medical slides, and electron microscope film, but they are all rather expensive.

Another important feature is the scanner's capacity for bulk scanning. For that, you load the scanner with material, configure the scan, and let it run unattended. It is almost impossible to digitize an extensive image archive without powerful bulk processing. It is simply unrealistic to digitize several thousand slides one by one. The slide feeder SF-210 lets you run stacks of up to 75 slides (with thin mounts) through the Nikon Coolscan 5000 without supervision. This vastly reduces your workload - as long as no slides jam.

A film scanner should support bulk scanning at least for negative strips. A motorized filmstrip adapter is now a standard piece of equipment. Here the mechanical quality of the scanner is important, so that the film is advanced precisely to the next frame after every scan. Bulk scanning requires high quality mechanical hardware. If the film transport or the slide feeder frequently jams and requires manual interaction, the feature becomes useless. Scanners such as the Reflecta DigitDia 4000 can scan entire slide trays even in their basic configuration.

IR Scratch removal is a must.

An up-to-date film scanner has to have a good implementation of scratch and dust removal. Although you can remove dust and scratches by software, too, this is way too time consuming and of inferior quality. Only use the software option when you have no other choice, i.e., when you scan b/w negatives.

Images Nikon

Nikon Coolscan 9000 ED for 35mm and medium format.

Nikon Coolscan 5000 ED with adapter for single slides.

The Expensive Nikon Way

Nikon has a different approach. Mounted slides, filmstrips, film rolls, slide stacks, and APS film can be scanned, but the scanner only comes with the slide mount and filmstrip adapters; any other adapter must be purchased separately. A film roll adapter alone costs far more than a basic film scanner. At that price, you get a slightly modified filmstrip adapter and a large film spool for winding up the film nicely.

Using a sophisticated adapter for this purpose is unnecessary, and it is not economical if you scan film rolls only occasionally. The Reflecta ProScan 4000 provides a much simpler solution: A second opening in the scanner lets you feed long filmstrips. Nikon offers no solution for scanning slides directly from the slide tray. You have to manually insert stacks of slides into the slide feeder. If you do not want to do that, you can get the DigitDia 4000 mentioned earlier, which is structurally identical with the Braun Multimag SlideScan 4000.

Coolscan 5000 ED with slide feeder for a maximum of 75 slides, depending on the mount thickness.

Images: Nikon

Left: Coolscan 5000 with filmstrip adapter for filmstrips with 2-6 frames.

Right: Coolscan 5000 with adapter for whole uncut filmrolls.

Scanner Light Sources

The scanner's light source greatly affects the scanning process. Traditional lamp technologies for scanners, such as halogen or fluorescent lamps, require a certain warm-up time before they can deliver a stable light. Meanwhile, newer technologies are taking over: for example, modern LED light sources provide stable lighting and therefore do not need any warm-up time. They also don't heat up, which is an advantage when scanning delicate originals.

For their current generation of scanners, Nikon and Konica-Minolta use LED lights. There are separate LEDs for each individual color channel. The disadvantage of LEDs is that they produce harsh lighting. With damaged originals, such as scratched black-and-white negatives, every scratch shows up prominently. This makes post-processing more time-consuming. A simple sheet of diffusing film placed in front of the scanner's light source can greatly reduce this unwanted effect. Alternatively, you can insert a special translucent plastic plate, such as the Scanhancer (see www.scanhancer.com).

While the Nikon image on the next page clearly needs touching up for the obvious scratches, the scratches on the Microtek scanned image are hardly noticeable. This shows how a scanner's design determines its performance. Harsh lighting adversely affects not only how scratches are handled but also how the film grain shows in the image. Soft lighting generates a much more pleasing image of the grain. Of course, having a scanner that does not emphasize the grain is better than performing grain reduction afterwards in software.

As demonstrated, you should be aware that, for technical reasons, one single scanner model cannot produce equally good results for all types of film.

▼ Reflecta ProScan 4000:
This scanner is suitable even for uncut film rolls.

▼ Reflecta DigitDia 4000:
This scanner can scan slides from their trays.

The harsh LED lighting of the Nikon Coolscan IV emphasizes scratches and makes them very visible. With color originals, you can compensate for this by using ICE, but black-and-white originals must be touched up manually.

The softer lighting from the fluorescent lamp of the Microtek ArtixScan 120tf makes the scratches barely visible and requires less touch-up work than the Nikon.

It's All in the Lens

Identifying your scanner's lamp type is easy. LED lamps are ready for action immediately after power-on, whereas older, cold-cathode fluorescent lamps need to warm up before they can be used.

A scanner lens projects the width of the film onto a CCD line sensor. This lens has to meet optical requirements typical of a top quality 35mm macro lens. State-of-the-art scanners have sophisticated lenses such as seven-element designs with exotic low-dispersion glass. Multi-format film scanners are even more demanding: the Coolscan 9000 ED has a lens with an astonishing 14 elements. Poor lenses are the main reason that scanners perform below their advertised resolution.

Ultimately you can always tell from the scans how much went into the design and manufacture of the film scanner. Please look at the test images on the included DVD. You will find big differences between scanners that have the same nominal specifications. A good film scanner can read enough information off 35mm film to enable high-quality 11" × 17" enlargements, provided that the source slide is of excellent quality. A good scan makes the weak points of the slide unmistakably clear, as it is like viewing the slide with an excellent magnifying glass. To get maximum picture quality on film, it is always recommended to use a tripod while taking pictures.

Just as in the case of conventional photography, the lens of a scanner is crucial for image quality. Only high-quality optics can generate a good image. You should consider this when comparing scanners. If the scanner's lens is of poor quality, good values for color depth and scanning speed are useless. In the end, only the final image results count. As manufacturing quality can vary over time, you should test your personal scanner with a USAF-1951 test-target. Even the best lens has to be centered properly during assembly to deliver good quality scans later on.

Images: Epson

Although most flatbed scanners used to be equipped with halogen or fluorescent lamps for a long time, this is slightly going to change. For example, the Epson Perfection V 500 Photo is using LED light

Double Stick Foam Tape

32mmx46mm Scanhancer

Double Stick Foam Tape

◀ Nikon slide mount adapter with installed Scanhancer.

2.2 Scanhancer

Originally Erik de Goederen developed the Scanhancer for the Minolta DiMAGE Scan Multi PRO. It is available at www.scanhancer.com and consists of a small polymer plate that is placed between the scanner lamp and the film. The idea behind the Scanhancer is to diffuse the scanner's light source, which de-emphasizes film scratches and also makes film grain appear more subdued. I have tested the Scanhancer on a Nikon Super Coolscan 5000 ED with slide mount adapter MA-21. My test results apply only conditionally to other scanner models. While the 5000 ED works fine with the Scanhancer, problems have been reported frequently with older models such as the 4000 ED. The Scanhancer was optimized for the DiMAGE Scan Multi PRO, which is still based on cold cathode fluorescent lamps. The image samples on the Scanhancer website were scanned using this scanner.

The above picture shows that the polymer plate was shortened for the MA-21. This enables calibration; otherwise it would cause error messages. Calibration does not happen directly through the Scanhancer but must be done separately. Also, after the conversion, the scanner no longer automatically detects when a slide is inserted. This automatic function, which starts a prescan after a slide is inserted, no longer works. Scanning times become noticeably longer with the Scanhancer, but in return scanning can be done without the processor-intensive GEM.

When starting the software with the Scanhancer, an error message pops up. However, just click to remove the error message and the scanner will continue to work. These are the unavoidable side effects experienced when modifying the scanner's hardware. For Nikon, the Scanhancer is not installed in the scanner body but in the slide mount adapter. Therefore, it is convenient to get a second MA-21 adapter for a quick change between scans with and without a diffuser.

Scanhancer can effectively subdue the harshness of an LED scanner lamp, which is especially useful when scanning black-and-white originals. Most scan software, however, is not geared to deal with Scanhancer, and using it often results in unwanted color shifts.

Scanhancer with Black-and-White Negatives

A common complaint about scanning black and white negatives, especially on standard Nikon scanners, is that the results are less than ideal. The harsh LED light points out every little scratch or dust particle. If you want to fix this, you have to spend a remarkable amount of time with Photoshop after the scan, but who wants to post-process every picture after the scan for an hour or more? So, I was curious to see whether using a Scanhancer would improve this obvious weakness of the LED technology. In my test, removal of scratches and dust particles worked properly after installing the scanhancer. However, as with all good things in life, there is a catch - there is a visible loss of contrast and fine details. Unprocessed black-and-white images not using the Scanhancer seem more contrasty and sharper. This becomes very obvious when comparing enlarged details. You must decide whether this is worth the time you save in post-processing.

Scanhancer with Color Slides

With color slides, scratches and dust are not a big problem thanks to ICE, even though ICE does affect sharpness in a way. But there is also the dreaded pepper-grain effect, which can somewhat spoil the fun of

▶ Initially the scanner is a little annoyed with the Scanhancer.

with Scanhancer   without Scanhancer

▶ Scanhancer suppresses many scratches and specs of dust, but at the expense of resolution and contrast.

scanning. When enlarging tight crops, quite often slide film looks like somebody sprinkled ground pepper over it. The film grain causes the little black dots, which are especially distracting in skin and sky areas. With a grain reduction filter such as GEM, this effect can be mostly eliminated. Depending on the amount of GEM correction, the image can get the plasticky look of digital photos, which can look quite strange. You have to decide whether this appeals to you; a finer representation of the grain would be better. This is where the Scanhancer is supposed to help. It does help, but not exactly as planned.

Dramatic Color Change

In the scanning program's preview window you can already see dramatically different colors. The colors become more muted and the image becomes generally darker with a noticeable color cast. The grain texture, however, is actually improved. It looks better than a standard scan. The Scanhancer suppresses many scratches and specs of dust, but at the expense of resolution and contrast. The image samples show slide film with relatively large grain. Do not expect any miracles from the Scanhancer, but it does do an acceptable job of smoothing the grain texture. With fine-grain films like Fuji Velvia, the Scanhancer smoothes the grain perfectly, but it alters the colors so badly that I chose to quit using the Scanhancer for color slides. Films with less saturated colors produce better results. But anyway: Color reproduction of a Nikon Coolscan 5000 is distorted by the Scanhancer, there is no doubt about it. As Scanhancer was developed originally for Minolta scanners, this may be an explanation for the reported symptoms.

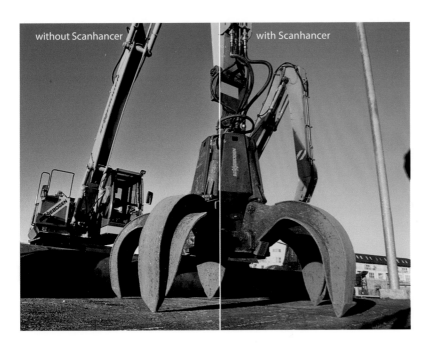

without Scanhancer   with Scanhancer

◄ When using the Scanhancer for color slides, the colors will be muted.

Scanhancer with Kodachrome Slides

Similar to normal color slides, Kodachrome slides will also exhibit a color cast with the Scanhancer. However, since Kodachrome slides have less vibrant colors than Fuji Velvia, the effect is not as bad. The color cast can be removed later in Photoshop. The Coolscan 5000 ED scanner has no ICE function specifically geared for Kodachrome. The only scanner with this function is the rather expensive 9000 ED. With the 5000 ED you can apply ICE to Kodachrome, but it removes image details that should actually be kept. In the sample image, you can see a cord that has been mostly shredded by ICE. This will not happen when using the Scanhancer. The conventional ICE (not optimized for Kodachrome) in combination with the Scanhancer can be recommended unconditionally. The scratch and dust removal through ICE no longer comes at the expense of details.

Scanhancer and Color Calibration

As already mentioned, the use of a Scanhancer causes a different color representation. It would therefore make sense to calibrate the scanner to compensate for this effect. This requires scanning software that supports IT8 scanner profiles. The images in the example have been scanned with a Nikon Coolscan IV and SilverFast Ai scanning software. Without correction, the Scanhancer cannot be used for color film. Most colors will be significantly off and the scan will have only experimental value, so this is not a practical solution for productive use. Theoretically, calibration should rectify this, since the calibration process compares a test chart to a reference file. However, you can see from the images below

This slide was scanned without Scanhancer and without ICE . Tiny specs of dust spoil the image.

Without Scanhancer but with ICE: The dust has disappeared mostly, but at the same time the cord has suffered badly.

With Scanhancer and ICE: The dust has disappeared and the details are still there.

that only the basic color cast will be eliminated. The color characteristic modified by the Scanhancer still differs from the reference scan. As can be seen from the individual color patches, tampering with the scanner hardware cannot simply be rectified with calibration. For certain color tones, there will still be deviations. Consequently, this solution is currently not recommended for color material. This is true at least for the Nikon scanners I tested; other scanners may be different.

Final Verdict

The final verdict on the Scanhancer is mixed: It can be effective when scanning black-and-white negatives, albeit with a slight reduction in contrast and sharpness. The disadvantages, however, outweigh the advantages when it comes to scanning color negatives and slides. The potential advantages cannot compensate for its often inaccurate color reproduction. The Scanhancer principle is still convincing, but there has to be more adaption for specific scanner types before this solution is ready for the end-user market.

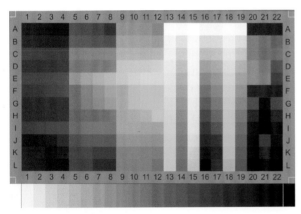

Without Scanhancer:
reference scan of an IT8 calibration slide.

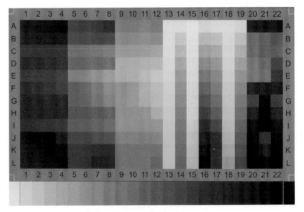

With Scanhancer and without re-calibration:
There is a major color shift.

With Scanhancer and re-calibrated:
The main color-cast is gone, but there are still some color shifts.

2.3 Interfaces: Connection Between Scanner and Computer

Data is transferred from the scanner's internal processing to the computer for storage. For this purpose, both scanner and computer must have a common external interface. The usual types of interface for scanners are SCSI, USB, and FireWire. Current models are offered almost exclusively with either USB 2.0 or FireWire. A PC can be upgraded with inexpensive expansion cards if the required interface is not included in the basic package.

SCSI

In the days before USB 2.0 and FireWire, SCSI was the fastest external interface. Older scanners still use SCSI, but modern units are equipped with the more user-friendly FireWire or USB 2.0. Unlike SCSI, they support plug-and-play.

USB 1.1 und USB 2.0

▼ Do not hook up USB 2.0 devices to older USB 1.1 hubs.

Image: D-Link

The Universal Serial Bus can connect practically any peripheral – keyboard, mouse, printer, scanner, and many other devices – to the computer, with the exception of the monitor. USB is a bus that allows for many devices to be connected in a tree topology, whereby USB hubs form the junctions for connections that branch out.

To utilize full bandwidth, connect the film scanner directly to a USB socket on the rear panel of the motherboard. If you place a USB hub in between, it could become a potential bottleneck for the high data transfer rates typical of film scanners. Theoretically, 127 USB peripherals can be connected to a PC, but they have to share the bandwidth of the bus. If you connect not only the scanner, but also printer, keyboard, mouse, and maybe even an external USB hard disk, the film scanner will not be able to use the full bandwidth of the interface. This can slow down the scan. There are two versions of the USB standard: USB 1.1 with a maximum bandwidth of 12 Mbit/s and USB 2.0 with a nominal bandwidth of 480 Mbit/s.

High Speed is Way Faster than Full Speed

Don't be fooled by the misleading terms "full speed" and "high speed" crawls at 12 Mbit/s. Newer PCs are equipped almost exclusively with USB 2.0, which is backward compatible. USB 2.0 devices can also operate in USB 1.1 mode. One side effect is that a single USB 1.1 device will slow down an entire USB 2.0 channel to USB 1.1 speed. Current computers typically have four to six USB ports on their motherboards. You should dedicate one port just for the film scanner and avoid connecting any other devices between PC and scanner. USB supports Hot Plug and Play, which lets you plug and unplug devices while the computer is running.

Don't Restrain Your USB 2.0 Scanner

If your scanner supports USB 2.0, you should also connect it directly to a PC with a USB 2.0 port. You will slow it down unnecessarily by connecting it to a USB 1.1 port. This also applies if a USB 1.1 peripheral is connected to the same hub as your USB 2.0 scanner, or if the scanner is daisy-chained to a slower device. A common mistake is to install a USB 1.1 driver even though the motherboard supports USB 2.0. In the Windows Device Manager you can see whether the correct driver is installed. In some cases you may also have to check your BIOS settings to see whether USB 2.0 is enabled or not.

The computer no longer has to be switched off before connecting a new peripheral. Still, the scanner programs I know require the scanner to be turned on before launching the program. Not all USB cables support data transfer at the maximum speed of USB 2.0. To be on the safe side, it is best to use the original cable supplied with the scanner.

IEEE-1394b/FireWire

The IEEE 1394b standard is marketed under the names FireWire, i.Link, and Lynx. They are all based on the same technology. There are two different types of connectors: four contact and six contact. The difference between the two types is that the 6-pole connector also provides power for the bus.

With a maximum of 400 Mbit/s, FireWire supported high data transfer rates long before USB. FireWire is a mature technology. Despite that, the USB standard promoted by Intel has widely prevailed. New scanners are more likely to come with USB 2.0 rather than an IEEE 1394b interface.

The advantage of FireWire over USB is that devices cannot slow each other down. FireWire always gives each device full bandwidth, whereas devices on a USB bus have to share the bandwidth of a single data bus, which can lead to performance limitations. If, like the Epson Perfection V750 Pro, your scanner has a FireWire connection as well as USB, you should use FireWire. FireWire technology is definitely superior, even if USB's relative cheapness has given it a greater market share.

▼ A PCI extension card for upgrading to FireWire.

Image: Western Digital

2.4 Life After the Purchase

Learning Curve

Film scanners and their related software are complex. Just as you would not expect to learn how to use an SLR camera in one evening, you should allow sufficient time to get the hang of your scanner and its software. If you unpack your scanner and try to start off without basic knowledge, you will yield only moderate results. At least one weekend should be dedicated to the first steps. This time is needed to become familiar with the scanner and the software. Once you have familiarized yourself and figured out a workflow, scanning will be a breeze. If you change to another scanning program, you will have to learn the system over again. If you can't or don't want to invest that kind of time, you should consider giving your slides to a scanning service instead.

Which PC is Suitable?

There is no simple answer to the question of what PC will enable you to work comfortably with a film scanner. Apart from the scanner model, the answer depends on the scanning software and its activated options. For example, enabling ICE will roughly double the scan time. The manual for Nikon Scan version 4.0.0 specifies a 300 MHz Pentium with 512 MB RAM as a minimum requirement. So, I figured I was pretty well equipped for my first attempts at scanning with a 1.6 GHz AMD Athlon, 768 MB RAM, and a 120 GB hard disk.

Now I know that my system at that time was rather underpowered for a film scanner. The processor usage during scanning is maxed out almost continuously at 100 percent. The computer runs constantly at its limit, which greatly prolongs the scan times. Also, the hard disk fills up quickly when image files are upwards of 110 MB each. For scanning large numbers of images at a reasonable speed, you'll need a computer with at least the following configuration:

▸ An up-to-date PC is essential for a smooth operation.

- ▸ 3 GHz CPU (multi-core recommended)
- ▸ 1024 MB RAM
- ▸ Fast, large hard disk
- ▸ USB 2.0 or FireWire interface

The above is just a rough guideline, since the resources needed for a scan depend on many different factors. Even with a fast computer, the scan ties up almost all resources. Thanks to multi-tasking, it is possible to do some other work in parallel, but only very slowly. If you suspect that the processor is fully occupied during scanning, you can check the Windows Task Manager. The CPU usage is shown under the Performance tab in a graph. Anyway, since multi-core processors have entered the desktop market the situation has improved greatly. Even if your scanning program occupies all the resources of one CPU core, parallel tasks can be handled by another CPU core. Apart from a powerful, high-quality

Image: HP

computer, it is also worthwhile to invest in a high-quality monitor. Both CRTs and high-end LCD monitors are suitable for image processing. Since the monitor is the main user interface, you should not skimp here. Good image processing is not possible on a bad monitor.

Noise

The first time you use a film scanner, you may wonder what on earth it is doing with the film. The buzzing and screeching noises coming from the scanner do not bode well. Only when the scanner spits out the un-harmed original will you realize that this is actually the normal sound of operation! As far as I know, there is no film scanner on the market with a pleasantly low operating noise level.

Customer Support

Scanners are complicated machines, so occasionally hardware or software can malfunction. This applies to systems from all manufacturers equally. For that eventuality, good customer support is essential. You should consider only those manufacturers who continually maintain their products and issue bulletins about bugs and incompatibilities. For repairs, you will have to rely on qualified service centers.

The current crop of multi-core processors (Core2Duo, Quadcore, etc.) are predestined for use with scanning applications, as they make it possible to continue using your computer normally while running scans in the background. Even the most powerful multi-tasking single-core processors cannot run background processes reliably.

Image: Eizo

CG-221K from Eizo is a high-end monitor which can display the full Adobe RGB color space.

Before buying, check for the following on the Internet:
▸ Is there a support hotline? Can it actually be reached?
▸ Is there an FAQ listing where you can find help?
▸ Is there an updated, fully-searchable knowledgebase?
▸ Does the website offer updated drivers, software, and firmware updates for download?
▸ Is there a service center or a manufacturer-certified repair shop in the United States?

You can Google the Internet for a scanner model and related key words to find user reports. They are often very subjective, but you can easily find out whether a specific model has many problems. Customer support is costly for manufacturers, and only rarely can the costs be passed on to the consumer. Therefore, many companies provide only very limited support or a toll number for telephone support. The manufacturers try to filter out the majority of the inquiries with FAQs and a knowledge base. This is an acceptable solution if they are user-friendly and well maintained.

Firmware Updates

Film scanners are controlled by firmware. This is software that handles the basic functions inside the scanner. The user cannot see the firmware, but rather interacts with the scanner via other applications, such as the scanner software. The scan program can only work optimally if the firmware is up-to-date. Every now and then, there are updates for the firmware to correct known problems. It is recommended to check the manufacturer's website regularly for such firmware updates and to install them. With Nikon Scan, the scanner's actual firmware version is displayed in the title bar of the scan window.

▸ Firmware updates for Nikon scanners are available at nikonusa.com.

Software Support

Most scanners are shipped with a scan program. The difference in program quality among scanners is quite considerable. Although the scan program from Nikon has a very good reputation, there are good reasons for using third-party software such as VueScan or SilverFast instead.

Standard products such as SilverFast and VueScan support almost all popular scanners, but surprises are possible since they do not adapt equally well to all scanner models. There actually is a difference as to whether SilverFast/VueScan is used with scanner model A or B. One look at the compatibility list will clarify whether your scanner is supported at all, but it will not tell you how well the program was implemented. Before you buy one of these programs, you should ask for a demonstration with a few sample scans.

Not even the manufacturer's own software may support all of its scanners. For example, Nikon Scan may not support all Nikon scanners. Apart from the current generation, it runs on only one previous generation. Older models require older software. Apart from Nikon, I have tested almost all current 35mm film scanners and a few flatbed scanners in researching this book. Be they HP, Epson, Microtek, Canon, Plustek, Reflecta, or Braun, they should do the customer a favor and include a current copy of either SilverFast or VueScan. In all cases, the software normally included is barely adequate.

▼ There is no substitute for VueScan or SilverFast. The manufacturers' included programs have lots of buttons, but only a few usable functions.

2.5 Wish List for Future Film Scanners

The current generation of scanners is already very good as far as resolution and image quality are concerned. However, there are areas where future technical advances will hopefully make the user's life easier. As there are often misleading ideas about the capabilities of the units, it is important to address a few points. I started this list in 2004 and was quite confident that we would see a new generation of high-quality scanners soon. Unfortunately, it is now 2009 and no major changes happened in the meantime! Anyway, I think it is important to point out the technical limitations of current scanning devices.

At the time of writing (January 2009), it is not worth waiting for new generation scanners to hit the market. Scanner manufacturers are not funding serious development, and we do not expect to see any significant technological advances in the near future. In contrast, even Nikon – the only residual manufacturer for high-quality desktop scanners – has announced it will completely stop its scanner production soon.

Scanning Speed

For many scanners, the speed of the scanning still leaves room for improvement. The norm is several minutes per image, which obviously slows down work. Faster scanning would be preferable. Scanning times of around one minute per image with image corrections turned on would clearly make scanning more enjoyable. With the current state of scanner technology, one weekend will not be enough to scan an archive of a few hundred images. Several weeks would be more realistic.

The fastest desktop film scanner I know is the Nikon Super Coolscan 5000 ED. Hooked up to a powerful computer, it allows for speedy work. Scanning a slide with activated ICE takes around one minute, which is quite acceptable. However, GEM and ROC further increase this time. Other scanner models are substantially slower. Scanning times of up to nine minutes for one image with ICE correction surely will not allow you to work efficiently.

Dust and Scratch Removal for Black-and-White

Once you have learned to appreciate the benefits of ICE, you will no longer want to do without it. Without much user interaction; slides will appear free of dust and scratches. Unfortunately, there is still no process supported by scanner manufacturers that also works for traditional black-and-white film. Especially with high-quality scanners, black-and-white negatives produce less than ideal results. The soft lens of a bad scanner makes some of the smaller blemishes disappear; yet, a good scanner scans not only the film but also every grain of dust on it and even the tiniest scratch with merciless precision.

Depending on the type of light source and the quality of the scanner lens, this effect will be more or less pronounced. You can remove the dust and scratches with software, but it is very tedious and cannot attain the results of ICE for color film. Scanhancer is a good tool to generally suppress dust and scratches. However, there is still no sound solution. Since all scanner manufacturers have frozen their development budgets, this problem will remain for the foreseeable future.

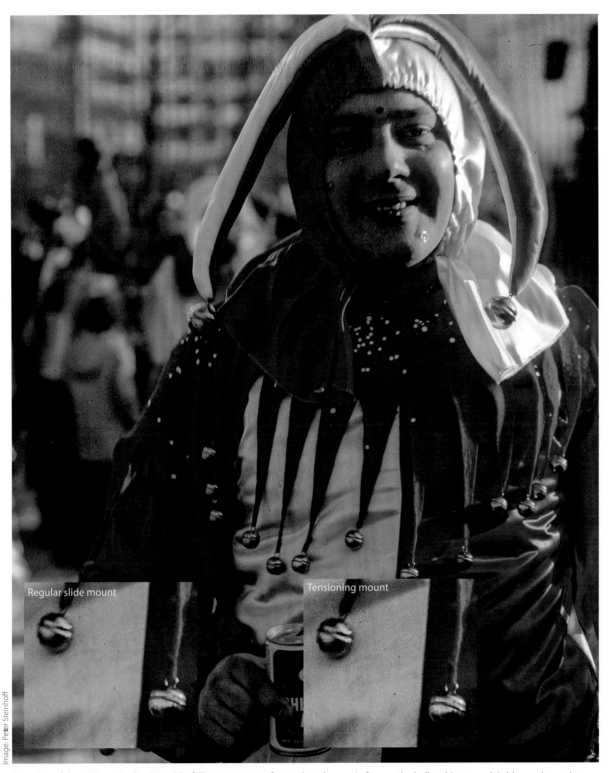

Snapshot of the 1980 carnival in Düsseldorf. The scanner was focused on the eye. Left crop: the bell on his stomach is blurry when using a regular slide mount. Right crop: The same image detail looks clearly sharper with a tensioning mount.

Depth of Field

Scan softness can have various causes, including the original itself. You can best check the sharpness of your original using a loupe and a light box. Scanners sometimes simply focus incorrectly, and warps and bends in the original are also often the cause of the problem.

In my scans, the depth of field issue has often reared its ugly head. The color of the scan is fine, but the sharpness is often limited to certain areas of the image. This problem always happens with warped originals and can be reduced only with tensioning mounts. In such cases, one has to choose whether to focus on the image center or the edge. A certain loss of sharpness toward the corners of the image is a natural characteristic of any camera lens, so the film is already less sharp there. Unfortunately, the limited depth of field of the scanner lens further adds to the softness in the picture, and ultimately this leads to visibly blurry portions in the scanned image. This problem varies with different manufacturers. Nikon scanners are infamous for their shallow depth of field.

Originals are not always sufficiently flat to be scanned equally sharp from center to edge. The images on the previous page demonstrate this nicely. At first, a curved slide was scanned. It was not possible to maintain even sharpness across the entire image; the eyes and the bell on the jester's stomach cannot be sharp at the same time. Then the slide was remounted into a Wess slide mount with a tensioning feature, which pulls the film reasonably flat. With reasonably flat film, the sharpness extends across the entire area. In both cases, the focus point was the left eye of the jester. ICE was deliberately disabled, since it causes slight blurring.

Tensioning slide frames is too troublesome and expensive for a large number of slides. The Wess website has not been accessible for some time now, so the company may be out of business already, but you will still be able to find some of their frames at www.filmscanner.info. Furthermore, it is not common to frame negatives, although it is possible for the purpose of scanning. Keep in mind that once the negatives are cut, they are no longer accepted at photo lab chains.

Nikon has not found an ideal solution for curved filmstrips either. In my opinion, Nikon's FH-3 filmstrip holder still has plenty of room for improvement. Wavy filmstrips become a little less wavy, but a proper flatness is not achieved. Unfortunately, I don't know of any better film adapters from other makers. In the twenty-first century, it ought to be possible to build a film holder that can pull curved film flat.

Ergonomics

Anyone who has used a film scanner before will understand why the issue of ergonomics is brought up here. Many models produce buzzing and chattering noises that can be unpleasantly loud. You have to be very indifferent to noise if you want to run a scanner in your living room and do other things at the same time. Flatbed scanners are better in that respect; their noise level is usually more agreeable.

Scanning Film

Scanned images cannot turn out better than the original film which was scanned. Every defect on the film will be recorded by the scanner as an image detail. It can be frustrating to start with an image that appears satisfactory as a 4" × 6" print, only to find noticeable image defects when the same image is viewed on the monitor. One must check whether such defects are caused by the scan or originate from the film. Quite often a poor original is to blame. If an image is scanned at 4000 dpi and viewed at full size, even the smallest defects become visible. The film should be checked with a good loupe. In many cases the film is fine, but the scanner is poorly focused. Even the best scanning method cannot compensate for a blurry original on film.

3

Contents

3.1 Handling Film

Curled, Wavy Film

Always handle your slides and neg-atives with care. You cannot fully compensate for defects in the scan-ning process.

Curled or wavy filmstrips are a problem that is commonly encoun-tered when scanning. A good scan requires a flat original. High-end film scanners have a very shallow depth of field – a curvature of just 1 mm can cause a blurry scan. Some film holders manage to minimize this curvature, but they are generally clumsy. Whenever possible, it is best to start with flat film stock.

Various factors determine the flatness of films. Non-curling layers in the film base can reduce excessive curvature of the material. However, a certain amount of curvature is required, allowing the film to stay in good contact with the film pressure plate inside the camera, and to avoid arching toward the shutter. Unfortunately, there is no binding standard for this concave curvature; manufacturers use their own guidelines.

Film processing and storage methods can also have a major influence on film flatness. Mistakes made in processing and storage can cause trouble later. For that reason, it does make a difference whether films are developed by a professional lab or by a discount processing service. If the film does not get dried according to instructions for that particular type of film, it will curl more strongly.

Though standard processes are used for all common color films, the results of processing can still vary greatly. The key processing factors for each lab are in how well they manage their quality control and how close-ly they follow the film manufacturers' processing guidelines. Ideally, the result of high-quality processing is that the film is flat after processing and can be scanned without difficulty.

How to Deal with Curled Film

There are limits as to how much film curvature can be corrected. The film base tends to return to its original shape by virtue of a memory effect. The simplest method to correct curvature is to store the negatives pressed between book pages. (Note: If this method is used, the negatives should be kept in their protective sleeves to avoid direct contact with the book's ink.) Also, a press could be used for storage. If the film is badly warped, it may help if the film is washed and dried once more. Unfortunately, there is no guarantee that any of these methods will truly keep the film flat.

Apart from lab processing, film storage is also a key factor in dealing with film curling. Storing films rolled up in a film canister may make it difficult for them to ever be flat enough for scanning. Often a flat filmstrip is fed into the scanner's film gate only to reemerge curved and curled lengthwise, like a pig's tail. This is most likely related to mechani-cal stress during the motorized film advance.

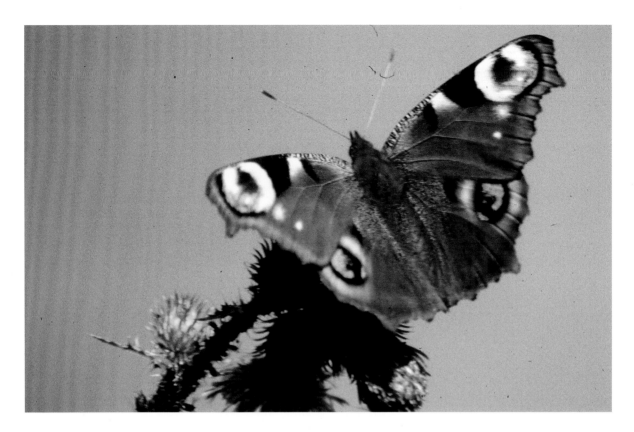

Whatever the reason, labs do not always deliver flat film stock. Typically, a mechanical fix is required to produce flat film. The best solution is to use slide mounts with a tensioning feature, such as the mounts available from Wess. These keep the film flat even when the negatives are very badly curved. However, to use slide mounts, the negative strip must be cut; once this is done, the negative can no longer be processed at most large labs (for example, when ordering reprints).

▲ High-grade film scanners capture even the smallest fibers or dust particles. Often, such defects can be seen even at standard magnification, as is the case here on the antenna.

Physically Removing Dust

Before scanning, all dust and dirt particles should be removed from the film. The dirtier the original, the worse the resulting scan. The better the scanner, the more clearly even minute contaminations will be recorded. Although these can be removed to a certain degree with Digital ICE or FARE, it is better to first clean the film properly. If a large speck of dust obstructs a part of the image on the film, then even the best algorithm cannot adequately reconstruct the image. When cleaning with a normal cloth, there is a danger of statically charging the filmstrip. This has the undesirable effect of attracting even more dust to the strip. It is best to clean filmstrips with an aerosol dust remover or a blower. Avoid using air from a compressor, since the air may contain traces of oil. Blowing away the dust with your mouth is also not advised, since breath contains too much moisture, which can lead to Newton rings in the scan.

If the film is statically charged, even cleaning with an aerosol dust remover will not be very successful as the film will attract new dust particles from the surroundings; having just cleaned the film makes almost no difference. If negatives are kept in plastic sleeves, they should not be pulled out of the sleeves quickly. Pulling them out quickly generates a static charge and causes more dust to gather on the film surface.

Dust Reduction with Agfa Films

For several years Agfa has included anti-static layers in their films. Because of this, Agfa films stay cleaner than conventional films. Dust can still deposit on Agfa films, of course, but without static charge dust will not be actively attracted to the film base.

Inserting Film Correctly

When film is inserted into the scanner, it is important that the film be oriented correctly. Film has two different sides. The emulsion side contains the image information. The emulsion is a thin layer deposited onto a high-gloss film base. It is the emulsion side which must be scanned for best results. Of course, film is transparent – if it is scanned the wrong way, there will be only a minor loss in quality.

Finding the Emulsion Side

You can identify the emulsion side of film by holding the film at a steep angle against a strong light. The film base is glossy and smooth, but the emulsion side shows relief lines from the image details. Slides have very visible relief lines. For negatives, however, it is not so obvious; the less glossy side is the emulsion side. Every filmstrip also features frame numbers which can help you determine the emulsion side: if the numbers are mirrored, then the emulsion side is facing you. (Note: for best scanning results, the scanner should read the emulsion side.)

Handling Film Strips

Both negative and slide film are delicate and subject to damage due to dust and scratches. If you want to avoid loss of picture information, you have to store your filmstrips carefully.

It is very important that film be handled and stored with proper care. Slides are easy to handle as they can be grabbed by the wide frame of the mount. Negatives are more delicate and difficult to handle. Avoid touching negatives with your bare hands. Fingerprints damage the film

Scanning with Correct Film Orientation

For an upright standing Nikon scanner, the emulsion side of the film must face downwards. If you can read the frame numbers properly from the top, the film is oriented correctly.

surface and are tough to remove from scanned image files afterward. It is better to remove fingerprints beforehand with film cleaners such as Tetenal and Hama. You can also safely handle negatives with thin cotton gloves, or use tweezers with slanted tips to remove the strips from their archival sleeves without damaging them. These tweezers are relatively inexpensive and can be found in shops catering to stamp collectors. It is recommended that negatives be stored in either glassine (matte) sleeves or archival-quality clear sleeves. Be sure to keep your film archive free of light, dust, humidity, and heat.

Cotton gloves protect film from scratching, but themselves produce fine fluff which can show up on scans. It is always best to handle negatives and slides by their edges, even if you are using gloves.

Destroy Originals after Scan?

After a successful scan, the original file is no longer needed; yet, it is a good idea to keep your slides and negatives on hand. Future generations of scanners may produce better results than currently available scanners; and in case of data loss, as a last resort, you can always go back to the original film and rescan your images. Data formats and storage media are constantly changing and must be converted on a regular basis to stay compatible with current systems. In the case of 35mm film, the situation is somewhat better; even negatives as old as 50 years can be used for printing or scanning without problems.

It can make sense to rescan certain images. Results of scanning depend, among other things, on the skill of the user. That is, if you gain more experience and rescan an original at a later date, you may get a better quality image. The best recommendation, therefore, is to never discard your analog originals (although it does make sense to sort out and discard any bad shots to reduce your archive).

Image: Gepe

Clean and compressed air is the safest option to get rid of dust on your filmstrips.

3.2 Film Types

Color Negatives

Color negatives are not very troublesome for shooting or for scanning. That is the reason why the overwhelming majority of photos taken in the past have been shot with negative film. For the purpose of scanning, these films have advantages, too. Even an inexpensive scanner can yield acceptable results when color images are scanned. One disadvantage is that the film base for color negatives is susceptible to scratches. You can get good scans from color negatives only if you use a scanner with good hardware-based scratch removal.

When looking at the inverted filmstrip with the orange masking, it is difficult to predict how the final image will appear. Even proper exposure and sharpness are difficult to judge. This shows that it is not easy to sort out unusable shots. The characteristics of film emulsions are as manifold as the offerings of color negative films. The data read by the scanner must be interpreted accordingly, and the result of scanning depends greatly on the settings of the scanning software. Matching the original film, in this case, is much more difficult than it is for slides. Up to the moment there is no calibration possible for color negatives, i.e., USAF-1951 only works with slides. Additionally, the scanning process for negatives takes longer than for slides because conversion from a negative to a positive image requires more processing effort.

The color correction of scans can be automated partly in scanning software, which supports film profiles, but it is not very reliable. Silver-Fast and VueScan ship with a large number of profiles for common film types. Nikon Scan has only one standard profile for negatives.

Color Slides

In a purely analog workflow, slides offer the maximum obtainable image quality. Slides offer much more vivid images than color negatives; their contrast range is clearly higher. However, slides demand accurate exposure from the photographer as there is little margin for error. Even a small exposure error affects the image. In comparison, one has much more leeway with exposure when using color negative film. The processing lab has little impact on the image result when working with slides. With negatives, it is a different matter. Slides are superior to negatives because they have finer grain, more contrast, better color depth, and a much better density range. For scanning, slides are more demanding than negatives. For example, scanning slides using low-end scanners often leaves much to be desired, since such scanners cannot typically handle the wide range of colors, contrast, and density inherent in the slides. On the other hand, a great advantage of scanning slides is that the image on the monitor can be compared directly with the original. Scanning slides is also faster than scanning negatives since there is no need for processor-intensive image conversion when slides are scanned.

◀ Coolscan IV: Scanned with a setting for standard slide film, this Kodachrome slide has a blue color cast.

◀ Coolscan IV: The same slide, but scanned with a setting for Kodachrome, this slide has neutral colors.

◀ Setting a Coolscan 5000 to Kodachrome produces the most vivid colors.

Image: Peter Steinhoff, ca. 1970

Kodachrome Slides

Kodachrome films are color transparency films that have been on the market for more than half a century. They are not developed with the common E-6 process, but rather with the special K-14 process from Kodak. There are still good reasons for professional photographers to use Kodachrome:

▸ With proper storage (in darkness), Kodachrome colors can remain stable over decades. However, under projection, they tend to fade faster than E-6 slides.

▸ Colors within the Kodachrome family of films are consistent, which is an advantage when different films are mixed in a slide show.

▸ Kodachrome films have very fine grain.

These films cannot be scanned in the same way as the common E-6 color slide films. They can be scanned like normal color slides, but with only modest results. You will get a good scan only if the scanning software offers a special setting for Kodachrome. If you scan Kodachrome with an E-6 setting, the images will have a bluish cast, and generally the colors will be less than satisfactory. Hardware-based scratch removal systems such as ICE or FARE, which were developed for E-6 and C-41 films, work for Kodachrome, but only partially. If you use scratch removal anyway, expect some loss in image detail.

Black and White Negatives

In general, scanning black and white negatives is problematic. As with color negatives, black and white negatives suffer from dust and scratches; but in this case, hardware-based scratch removal does not work at all. Neither ICE nor FARE function properly for black and white negatives; they actually degrade image quality. However, ICE and FARE work for black and white films based on the C-41 developing process. These films are closer in technology to color film; therefore, the infrared beam used in ICE and FARE can scan the surface for dust and scratches. These films are called "chromogenic" color films, such as the Ilford XP2 Super, Kodak BW400CN, Konica Monochrome VX 400, Tura BW-C41, and Fujifilm Nexia Sepia. Traditional black and white films such as the Kodak T-MAX family are silver-based.

> ## Improving Flatness for Negative Films
>
> If a negative strip cannot be scanned properly, it can be cut and mounted. Tensioning slide mounts produce flatness that is superior to typical film holders. Note that once the strip is cut up, labs will no longer handle it. This is basically not a problem, however, since you will have a high-quality image file once the negative is scanned.

The Ideal Film for Scanning

Black and White Negatives? ▸ Opinions are divided about the ideal film for scanning, but it is certainly not black and white negative film. Black and white does not allow dust and scratch removal with ICE. In a purely analog workflow, the qualities of good black and white films remain unsurpassed, but the annoying problem of scratch removal makes it difficult to take advantage of those qualities in digital processing. High-quality scanners reproduce even the smallest scratch, which can actually degrade a high-resolution image. Theoretically, you can use image-editing software to touch up scratches, but doing this is very tedious and time-consuming and is not practical for scanning a large number of images.

Color Negatives? ▸ The advantages of using color negative film for scanning are wide exposure latitude and limited contrast range. Even with a low-cost scanner, quite passable scans are possible with color negatives. It can be difficult to visually match the scanned image with the negative. Orange masking makes it difficult to judge image sharpness, and it takes a lot of experience to guess the colors hidden underneath. Still, with matching film profiles, you can easily produce scans with good colors. Negative film is more prone to scratches than slide film, but in most cases these scratches are handled automatically with ICE. For average quality needs, color negative film is a good compromise.

Color Slides! ▸ Slides have more vivid and vibrant colors and a wider density range than negatives. Only superior film scanners are able to read such image information. A cheap scanner won't make you happy in the long term, and it is always better to invest a little more in your equipment at the outset. Matching a scanned image with a slide is straightforward. Correct exposure is more critical for slide film and requires more caution during shooting. Slide film also has the advantage that it is less sensitive to scratching than negative material. Scanning slides is faster than negative scanning, as it does not involve the processor-hungry inversion and de-masking steps. You also don't need to use individual film profiles in order to render colors accurately. If you know you are going to scan your results at some point, you should always use slide film in preference to negative film, as this will produce the best results—especially if you have use of a high-quality scanner. Even the best scans, however, can never be as vivid as the image produced when the original slide is projected onto a screen. The best stock for good scans is a slide film with fine grain; the finer the better. While expensive drum scanners used by professionals can handle almost any film stock, desktop film scanners used by amateurs and semi-professionals are more limited. Before the digicam boom, large film manufacturers developed several films optimized for their ability to scan. If you are still shooting film and scanning by yourself, you should use film with this newer technology. The difference is obvious; with older film you will get inferior results.

Scratch Removal for Kodachrome
At this time, the Nikon Coolscan 9000 ED is the only film scanner offered specifically for Kodachrome with a version of ICE scratch removal.

▲ An old slide: the scan with the glass frame causes a loss of sharpness.

Granularity of Film

The familiar root-mean-square (rms) granularity values have been gradually replaced by print grain index. While rms granularity is an absolute value, print grain index varies with print magnification. There is no direct correlation between the two. An rms granularity of less than 10 and a print grain index of less than 25 (35mm film enlarged to 4″ × 6″ print size) represent a very fine grain.

Slide Mounts: Glass and Glassless

The type of slide mount has a big effect on the quality of the scanned image. Slide mounts with glass inserts are not suitable for scanning, and should be replaced before scanning in order to avoid significant loss of sharpness.

Glass Mounts

Slides mounted behind glass stay flat, which at first glance seems equally advantageous for both projection and scanning. With scanning it prevents popping of the slides and keeps the surface even. However, for scanning, the glass layer is an additional obstacle when it comes to image quality – even more so with anti-Newton glass mounts. If you want a good-quality scan of your glass-mounted slides, it will be necessary to remount your slides into glassless frames. This applies to single and double-sided glass frames. Unlike slides that are projected, slides will not "pop" in modern scanners since they do not heat up.

Glassless Mounts

Glassless mounts are ideal for scanning, since there are no trouble-some glass layers involved. In many cases, however, flatness of the film is a problem. Mounts are necessary to ensure good flatness. Machine-mounted slides from the lab are usually firmly fixed and do not buckle. Other mounts, such as those used with the Hama DSR system, do not offer any locators for the film. Wess has addressed this problem by developing several mounts with a tensioning feature; but they are all rather expensive and may be difficult to find. It is usually sufficient to purchase a single pack of mounts, as you will only be using them for scanning but not for actual archiving. As far as I know, these mounts are only available at www.filmscanner.info. The www.wess-diarahmen.de site is no longer online.

▲ Uncut transparency strips are perfect for batch scanning.

Images: Wikimedia Commons

◄ These slide frames are from 1942, but you can still handle them with every up-to-date film scanner.

Granularity of 35mm Kodak Film			
Film Designation and ISO Sensitivity	*RMS Granularity*	*Print Grain Index (enlarged to 4" × 6")*	*Remarks*
Color Negative			
Portra 160NC	n/a	36	Portrait film for natural colors
Portra 160VC	n/a	40	Portrait film for vivid colors
Portra 400NC	n/a	44	Natural colors and high speed
Portra 400VC	n/a	48	Vivid colors and high speed
Portra 800	n/a	40	Finer grain than Portra 400 films
Portra 100 T	n/a	33	For tungsten light (3200K)
Elite Color 200	n/a	32	General-purpose standard film
Elite Color 400	n/a	39	For action shots
Gold 100	n/a	45	Standard consumer grade film
Gold 200	n/a	47	Gradually replacing Gold 100 as standard film
Gold 400	n/a	49	Standard film for point-and-shoot zoom cameras
Color Transparency			
Elite Chrome 100	RMS 8	n/a	Universal fine grain film
Elite Chrome 100 Extra Color	RMS 11	n/a	Extra high color saturation
Elite Chrome 200	RMS 12	n/a	All-purpose film, fine grain for ISO 200
Elite Chrome 400	RMS 19	n/a	For low-light and action shots
Ektachrome E100G	RMS 8	n/a	Extremely fine grain
Ektachrome E100GX	RMS 8	n/a	For warmer colors
Ektachrome E100VS	RMS 11	n/a	Vivid, saturated colors
Ektachrome E200	RMS 12	n/a	Can be pushed to ISO 800
Kodachrome 64	RMS 10	n/a	Proven veteran of several decades; uses K-14 process. Color characteristic noticeably different from modern films.
Kodachrome 200	RMS 16	n/a	Quite a bit faster than Kodachrome 64; both films need to be processed in one of three remaining centralized labs in the world. Future support for 35mm is unclear; for Super8 and 16mm film, it was discontinued in 2006..
Black and White Negative			
T-Max 100	RMS 8	n/a	B&W film with extremely fine grain
T-Max 400	RMS 10	n/a	Still fine grain despite higher speed
T-Max 3200	RMS 18	n/a	The ISO-wonder: can be pushed to 25,000
Tri-X 400	RMS 17	n/a	Good for pushing, has characteristic grain
Tri-X 320	RMS 16	n/a	Slightly finer grain, otherwise like Tri-X 400
PX 125	RMS 10	n/a	Very sharp with fine grain
BW400CN	n/a	< 25	Extremely fine grain, uses C-41 process

Source: www.kodak.com

File Formats

The quality of scanned images depends greatly on the choice of the right scan file format and corresponding resolution. Which file format offers the best quality, is supported by most software manufacturers, and requires the least storage space? The following chapter will help you to decide on the ideal image file format and the best resolution for your purpose. Only formats relevant to film scanners will be discussed here. For good reasons, scanners currently tend to use RAW format. Just as digital photographers do, scanner users have to deal with incompatibilities between image and file formats.

Contents

4.1 Digital Negatives

Differences between Digital Cameras and Film Scanners

Software is a critical factor
The characteristics of a digital neg-ative depend not only on the scan-ner, but also on the software being used. A single scanner can produce totally different types of raw data files if used with different software packages.

RAW files are considered to be the "digital negatives" for digital cam-eras. High-end cameras offer the option of saving images in unpro-cessed RAW format. RAW is analogous to negatives or slides in the analog world. Each manufacturer uses its own RAW format, which can even vary among a manufacturer's different models.

With film scanners, it is a bit more complicated. Good scanning software also offers RAW as an option, but the type of RAW format de-pends on the combination of film scanner and scanning software. There is a big difference, for example, between generating RAW files from a Nikon scanner with Nikon Scan or with VueScan. Furthermore, if you have selected correction techniques such as Digital ICE before the scan, such corrections will be permanently applied to the image. Therefore, the meaning of a "digital negative" is not so clear when it comes to film scanners. The definition of a digital negative needs to be extended in comparison to digital cameras. The digital negative is the source file upon which all further processing steps are based. It must never be mod-ified unless the processing steps are reversible.

Conversion from Analog to Digital Through Scanning

Analog originals (slides or negatives) must go through a film scanner to become image files. This is where analog to digital conversion occurs. A file is generated from the analog original. Inevitably some image infor-mation gets lost in the process. During scanning, the goal is to extract as much image information as possible from the original and save it as a digital negative: the master file. Only then can the best image quality be achieved in post-processing. If any modifications to the image informa-tion contained in the digital negative are irreversible, the image would have to be rescanned to restore the original master file.

RAW comes closest to an ideal digital negative. If the scanning soft-ware does not allow you to save in RAW, you can use another lossless format such as TIFF as your digital negative. Whichever format is used, it is important that the full extent of the image information obtained with the scanner is preserved in the file.

Working with Digital Negatives

Image files are easily corrupted, and in most cases this happens acciden-tally by the user! Unlike an analog negative, the image file can be quickly and easily modified, and image information will be subsequently lost. If, for example, you crop a TIFF file and save it to a file with the same name, the cropped portions of the original image are gone. Nikon Capture Edi-tor handles this situation better than most with Nikon Electronic For-mat (NEF): the selected crop is saved in a configuration file as opposed to an image file without changing the underlying image. This allows one to revert to full image size later if necessary.

Conventional Image Editors

Conventional image editors offer less controlled access when working with an image file. For example, when working with Photoshop, one usually works on different copies of a file to avoid losses. With the right plug-in, Photoshop can open RAW files, including Nikon Camera NEF files, but modifications to an image can only be saved in a different format, such as TIFF. In this case, the digital negative - the master file – remains intact. If your source file is TIFF, on the other hand, the editor can overwrite the original with a modified version and permanently discard cropped borders. Therefore, modifications should be saved as copies, even though this costs additional storage space and can make it more difficult to keep track of versions.

Image Processing at RAW Level

It is more elegant to edit images at the RAW level, the way Nikon does it with the Capture NX2 Editor. All editing steps are saved to a configuration file and can be undone completely or individually. There is no need for an additional work-in-progress copy. There is also no risk of losing image information by editing. This is a truly nondestructive workflow. Any change is fully reversible.

However, editing at RAW level cannot yet offer all of the powerful functions available in Photoshop. Functions such as layers, retouching, or filters are generally not supported with RAW editing programs. Hence, RAW editors are less suited for creative image editing. And that did not really change with the introduction of Nikon Capture NX2 editor which has at least some of this features implemented. Although Photoshop does have functions similar to adjustment layers, image handling with Photoshop cannot compete with Capture NX2. For each correction (e.g., levels, curves, or color balance), a new layer must be created, and seemingly simple operations like reversible cropping are not possible. All in all, you have to use RAW editors and Photoshop in parallel to get full advantage for processing your pictures.

RAW: The Best Format for a Digital Negative

If you are not sure what format to use for a digital negative, scan in RAW. It can always be converted to another file format, but not the other way around. Be sure to keep the required converters together with your RAW files. Without converters, RAW files can not be opened. The type of RAW file produced depends on the combination of scanner and scanner software. Not all scanner programs support saving in RAW format and editing RAW files.

4.2 Color Depth in Image File Formats

An important criterion when choosing an image file format is the supported color depth. This determines the number of possible colors for each pixel (expressed in n bits). More bits mean more colors, but also increased storage requirements. The number of colors that can be displayed for a given color depth equals $2^{(\text{color depth in bits})}$.

The color depth of an image file is not permanently fixed and can be changed in the editor. This can easily lead to loss of image quality. Typical image editors work with either 8-bit or 16-bit color depth.

Adobe's Photoshop CS was the first image editing software to work almost exclusively with 16-bit color depth, although many of its functions only worked with 8-bit images. The CS2 and CS3 versions of the program have redressed the balance, and it is now possible to work in 16-bit without restrictions. Professional image processing has, however, moved on, and Photoshop CS3 already supports 32-bit color depth for HDR (High Dynamic Range) images. 32-bit color depth is currently not a significant factor in scanning processes, and most likely this situation will not change in the near future.

Images should always be scanned at the highest color depth available for the specific scanner. Only then will all the image information that was captured by the scanner be available for further processing.

1-Bit Color Depth

With a color depth of 1 bit, the computer has only a single bit for each pixel available, which translates to only two colors (i.e., black and white). This makes sense for text and line art only.

8-Bit Grayscale

This is a common standard for many file formats. Each pixel can have one out of $2^8 = 256$ levels.

8-Bit in RGB Color Space = 24-Bit Total

With a color depth of 1 byte, the computer gets 8 bits per color channel for each pixel. There are three color channels, one for each of the primary colors: red, green and blue.

Thus, the number of all displayable colors in this mode is $2^8 \times 2^8 \times 2^8 = 2^{24} = 16.777.216$ colors. Although each color channel has only 256 colors, the combination of the three channels makes over 16 million colors. It would be more correct to call it 24-bit color depth or 8 bits/channel, but it is often called 8-bit, which can be confusing.

◀ Two displayable colors: There is only black and white.

◀ Sixteen (not 16-bit!) different colors: The colors appear unnatural and subtle tones are missing.

◀ 16.7 million colors: The 8-bit color depth of the RGB color space delivers natural color gradations.

4.3 Important Image File Formats in the World of Scanning

JPEG – A compact format...

The JPEG format uses strong compression. This process reduces the RAW data from the scanner to a fraction of its original size. When opened, the file is decompressed, and the image expands in RAM. The small file size has many advantages: more files fit on a data carrier, and files can be easily sent over the Internet. Furthermore, JPEG is a popular standard and is supported by practically all image viewers and editors. But, it has one major flaw: with compression, some of the image quality gets lost permanently.

... but you lose picture information!

The degradation depends on the chosen compression ratio. The smallest file size requires the highest compression, which in turn causes the greatest loss of detail. A JPEG file in digital photography could be compared to a print from the lab in analog photography. The quality of the result depends on many factors, most of which are unfortunately out of your control. There are few possibilities to enhance the image afterwards; you have to accept it the way it is.

▼ In a direct comparison with a TIFF image, the blocking artifacts of heavy JPEG compression become clear. Here the sky shows highly visible banding due to excessive JPEG compression.

JPEG is poorly suited for processing in an image editor. If your goal is high image quality, JPEG should be used only at the very end of the image editing chain. JPEG only supports 8 bits per color channel, which is perfectly acceptable for the intended purpose of this format. However, as a file format for scans, JPEG is not ideal. It is fine for simple, quick scans; but if you want to postprocess your images, you will quickly run into limitations.

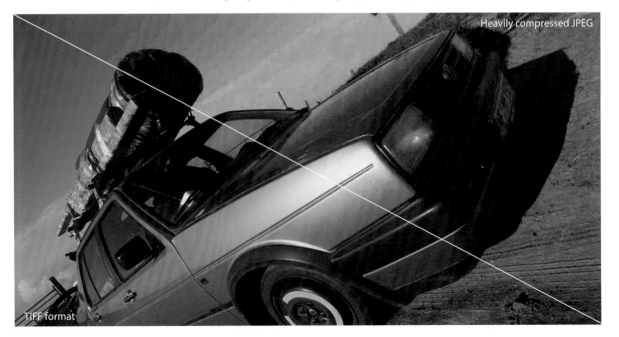

JPEG 2000 – Lossless Compression

Compared to TIFF or NEF, JPEG achieves good compression rates, but at the cost of detail and image quality. It has been more than 15 years since JPEG was invented, and its successor JPEG 2000, is waiting in the wings to replace it. JPEG 2000 allows high compression rates with less degradation, and it even offers a lossless compression option. You can choose between lossy and lossless compression. It mostly avoids loss of sharpness and some JPEG compression artifacts. While JPEG is not at all suitable as a digital negative, JPEG 2000 is basically acceptable. However, it is too early to recommend this new format since too few programs support it at present.

TIFF – Lossless Compression

TIFF is a general file format that accommodates many internal formats, several different compression methods, and different color depths ranging from 1-bit to 32-bit. 16-bit TIFF files from the scanner contain the maximum image information the scanner can produce. Since many scanners produce RAW data with 12-bit or 14-bit color depth, storage space is not fully utilized with a 16-bit TIFF file. A 12-bit scan must be saved as a 16-bit TIFF file. A corresponding RAW file would be more compact and use fewer system resources during editing and storage.

Most image editors support TIFF, but not always all of its many versions. In the analog world, a TIFF file would be comparable to a slide: with proper exposure, the ideal (analog) result is obtained. Carried over to the scanner world, exposure would correspond to the setting of the scanner software and the converter. Initially the scanner always generates RAW data. Afterwards, the data are converted to TIFF, which is a potential source of errors. For example, if you select excessive sharpening for the scan, it is almost impossible to fix it in TIFF. At RAW level with NEF this is no problem, but after converting to TIFF you lose the ability to correct the sharpening features.

Lossless compression of TIFF files is achieved in some image editors with Run Length Encoding (RLE), ZIP, or Lempel Ziv Welch algorithm (LZW) compression. Photoshop also offers JPEG-compressed TIFF, but a few scanning programs do not support it. For scanning, only the uncompressed 8-bit and 16-bit versions are of interest.

CMYK-TIFF

CMYK-TIFF is different from RGB-TIFF format in its use of the CMYK color space. Each pixel is described with four color values: Cyan, Magenta, Yellow, and Black (Key). Storage requirements for this format increase accordingly. This mode is needed only for prepress, and conversion from RGB to CMYK should only occur at the end of the job. There is no way to reverse the conversion from RGB to CMYK without loss, so the RGB original should always be kept.

Image formats in a nutshell
In spite of the many image formats available, it is relatively easy to choose the right one for your purposes. If you want to save storage space and you don't wish to post-process your images, JPEG is the right format for you. I nevertheless recommend that you scan your images in TIFF format, as this provides good image quality without compatibility problems. If you want to achieve the best possible image quality, and you have time and energy to spare, you should experiment with RAW data formats.

BMP (Windows) and PICT (Macintosh)

The BMP format is only relevant for Windows. The Mac counterpart is PICT. Both formats support neither 16-bit color depth nor color profiles. Therefore, they play no role in high-end image processing and should not be used for scanning.

RAW: Proprietary Image File Formats

RAW files are the internal data formats of particular scanners. They are not standardized – each manufacturer has its own format. Adobe's push to establish a standard with DNG format has not had overwhelming success – at least not with scanners. Neither Nikon, VueScan, nor SilverFast RAW files can be opened with Camera RAW and converted to DNG format. During scanning, the exact format of the RAW file will be determined by the combination of scanner and scanning software. RAW files from VueScan cannot be processed with Nikon Scan, and vice versa. Also, SilverFast HDR is not compatible with the other formats. (Note: SilverFast HDR has nothing to do with what Adobe means by "HDR".) In order to display and edit RAW files in conventional editors, the files must first be converted. The converter is the interface between RAW file and image editor.

Conversion is another source of potential degradation, since there are quality differences between converters. Conversion normally works only in one direction. Just one example: You open a digicam NEF-file with Photoshop (remember that NEF-Scanned is not supported by Adobe Camera Raw). Now the program is working not with the RAW file, but with a converted copy. Photoshop cannot save the edited image as a Nikon RAW file; it has to use alternatives such as TIFF. True RAW editors such as Nikon Capture Editor work differently where both editing and saving are performed at RAW level.

Meanwhile, there are RAW converters that have "write access" to RAW files, but only for camera RAW files. I don't know of any converter that can do that with scanner RAW files. This is a problem in that all RAW formats are proprietary and most applications need a suitable converter to open them. If you have a scanner and several different digital cameras, you will probably have to deal with at least three different RAW formats and the corresponding converters. You can avoid this problem if you work with JPEG or TIFF.

The Ugly Side of Proprietary Standards

Open standards simplify data exchange in many parts of the IT world, and it is always a benefit to the community in general when manufacturers open their standards to everyone. This is, however, often not the case with RAW data formats. Nikon keeps the specifications of NEF secret in order to promote sales not only of its digital cameras, but also of its own scanners and RAW processing software (Capture/Capture NX).

Difference between RAW and TIFF
The scanner generates TIFF files in a lossless conversion from RAW data. Therefore, it is a first-class file format. The difference between TIFF and RAW is that all the image corrections from the scanner program and converter are permanently applied to the TIFF file; the unmodified data from the scanner is only available with RAW.

This practice guarantees the manufacturer short-term turnover, but it artificially restricts the customer in his/her choice of software. Third-party manufacturers who want to construct NEF editors have to obtain their knowledge slowly and painfully using reverse engineering. Such practices are simply not worth the effort for less well-known NEF variations such as NEF Scanned, which explains the current lack of support for this particular format.

NEF – RAW Files by Nikon

NEF is a proprietary RAW format developed by Nikon. This format is a bit ambiguous. Files with the ending .nef can belong to one of four major categories, all of which are different. Nikon has created a seamless workflow for NEF files with its own programs Nikon Scan, Nikon View, and Nikon Capture NX2 Editor. The following chart shows the substantial differences in file size between uncompressed NEF/TIFF files and highly compressed JPEG files. As you can see, you can save a remarkabel amount of storage space, if you prefer JPEG over TIFF or RAW. But always remember: you have to pay the price through loss of picture information.

Poor support for NEF scan files

These days, there are many programs that support the Nikon NEF image file format, although this largely applies to files that were created using digital cameras. NEF scan file support is still something of a rarity. Adobe Bridge, for example, can preview a NEF scan file, but Photoshop cannot open it. To open and process NEF scan files, you need either the scan program itself or a Nikon image processor.

File Sizes of 35mm Scan @ 4000 dpi (Resolution 5292 × 3509 Pixel)	
JPEG-RGB	
Highest compression	626 KB
High compression	1 385 KB
Best compromise	2 269 KB
High quality	4 664 KB
Maximum quality	20 111 KB
TIFF-RGB	
8-Bit	55 728 KB
16-Bit	111 159 KB
NEF-RGB	
8-Bit	56 365 KB
12-Bit	84 548 KB
16-Bit	112 730 KB

VueScan RAW Files

VueScan is another program that can save in RAW format. Unlike Nikon, VueScan did not create a new RAW format, but rather uses special TIFF files. To use this format, you have to select the output option RAW file. These RAW files can only be properly displayed and edited with VueScan. The scanning program also works as a RAW editor. Unlike data in the Nikon NEF format, these are true scanner RAW data; no image corrections are applied to the file. This allows you to perform dust and scratch removal after scanning.

SilverFast RAW Files

SilverFast's own HDR files are also special TIFF files. Although HDR files are similar to the VueScan RAW files, the two formats are not compatible. HDR files generated by the scanner software SilverFast Ai are best edited in SilverFast HDR. SilverFast RAW is less consistent than VueScan, because ICE is applied to the data. All other corrections can still be done afterwards.

RAW Files in Detail: Nikon's NEF Format

Speaking about the NEF-format is quite tricky, because such a thing as a single Nikon NEF format simply does not exist! There are four major categories of NEF, and some of these even have dozens of subcategories. In the following section you will get a short introduction to help you understand why these files are all NEFs, but still can be very different in handling and processing.

NEF I: Original NEF

Uncompressed NEF files created by Nikon digital cameras are called Original NEF files. They can vary from camera to camera, which means that an editor like Capture NX2 has to be updated in most cases when new cameras hit the market.

NEF II: NEF Compressed

NEF Compressed refers to compressed NEF files from Nikon top-end digital cameras. The aforementioned Original NEF offers maximum image quality. Unfortunately, these files are big and quickly fill up expensive memory cards. Also, the writing cycles in the camera are relatively long, which can slow down continuous shooting. Good cameras can continuously shoot in the much smaller JPEG format without interruption until the memory card is full. When shooting NEF, the buffer fills up quickly and the shooting has to pause. Nikon has addressed this problem with the Compressed NEF format. The files are about half the size of the Original NEF files, and according to Nikon there is no visible difference between them and Original NEF. But, it is obviously not an entirely lossless compression. So far only a few camera models can generate Compressed NEF files. This format is not relevant for scanning.

NEF III: Scanned NEF

NEF III files are RAW data files generated by a Nikon scanner and saved by the program Nikon Scan. The software lets you select a color depth of 8-, 12-, 14- or 16-bit. Scanned NEF is a lossless format. With Nikon Scan several types of changes can be made to the Scanned NEF after the scan, such as the settings for brightness and contrast. Using this format affords image editing at RAW level with all its advantages.

Nikon Scan 4 can edit any type of file it has generated. However, postprocessing scanned NEF files with Nikon View or Nikon Capture Editor will change the files irreversibly. Apart from this minor nuisance, the NEF format comes closest to the ideal of a digital negative among all the formats introduced here. For an analogy in the world of analog photography, it is like a color negative that you can develop in your own lab. You have control over all key parameters and still can tweak it to a certain degree afterwards. This is the advantage NEF has over TIFF.

Strictly speaking, Scanned NEF files are not real RAW files; image corrections like Digital ICE, ROC, GEM, and DEE are applied to the data during scanning. If you want to change any of the settings, you'll have to rescan. In a true RAW file, all corrections are still possible afterwards without having to rescan. Unlike with VueScan, where you always have to create an additional image file from the RAW files first, you can open, edit, and save NEF files directly. Therefore, with NEF a single file is in many cases sufficient; image and negative are one. All changes are saved in configuration files in addition to the image data and thus can be reversed if necessary.

NEF IV: Converted NEF

Existing JPEG or TIFF files can still be converted to NEF files. All advantages of editing at RAW level are available, but you will not get real RAW data. Still, Converted NEF is a good way to protect files against accidental changes.

Consider the following example: you scan a slide as TIFF, whereby the scanner extracts the image information from the slide in RAW form. This RAW data gets converted, which means the converter's changes and manual changes – if any – will be firmly applied to the TIFF file. This TIFF can be turned into a Converted NEF. This Converted NEF file does not contain the original RAW data, but merely the image information of the TIFF file. The original RAW data was lost in the first conversion.

Scanned NEF Files: Post-Processing Can Cause Crash!

This example demonstrates a reproducible way to crash your scanner software: Take a scanned Image in NEF format and rotate it with the editor, which comes with the free Nikon View (not to be confused with Nikon Capture Editor, which you have to buy). Save the image. Now any attempt to open this file in Nikon Scan will crash the program; you can only open it with Nikon View, Nikon Editor, or Nikon Capture Editor.

Scanned NEF and Grayscale

Nikon Scan offers the choice of scanning in RGB or grayscale mode. Grayscale saves storage space, which makes it an obvious choice for scanning b/w negatives. B/W negatives can always be scanned in color, but then the images contain the color cast of the original. Neither Nikon View nor Nikon Capture Editor can open the grayscale scans since they do not support this color space. Editing at RAW level is only possible with Nikon Scan. Consequently, image editors that use the NEF plug-ins packaged with Nikon View also cannot open grayscale files. All in all it is the best choice to scan scan black and white negatives as NEF-RGB. You have to remove the color cast during post-processing.

LCH-Editor in Nikon Scan: The unmodified Chroma curve of the initial RGB file.

After pulling down the Chroma curve in Nikon Scan, all colors have been reduced to grayscale.

4.4 The Right Image Size and Resolution

Scanning film is normally done at the maximum optical resolution of the scanner. Even a scanner with 2900 dpi can produce 4100 × 2700 pixels from a 35mm negative. At 48 MB, the resulting file in NEF format is rather unwieldy and too large for many uses, such as sending it by email. The resolution should not be changed in the scan file (your digital negative), but for certain purposes it is unavoidable to have copies with lower resolutions.

Absolute Resolution, Relative Resolution, and Output Size

For printing, the image resolution is expressed in dpi. A resolution of 100 dpi means that 1 inch of paper width will depict 100 picture elements. This is a relative resolution, since the number of pixels relates to the physical print size. It can be changed at will without quality loss (just make sure to uncheck the option "Resample Image" in Photoshop, which by default is on). More important is the absolute resolution, which is expressed in pixels. For instance, reducing the absolute resolution will result in loss of detail and should never be done in the digital negative.

The absolute resolution of a desktop wallpaper might be 1024 × 768 pixels. From the absolute resolution and the relative resolution, the output size in inches can be calculated. Most viewers and editors calculate this automatically for you. In Adobe Photoshop, you can define each parameter separately, and the program will adjust the other parameters accordingly.

The output size of an image depends on the resolution of the output device. In our case the output devices are monitors, home printers, or the photo lab printer. To stay with the example of desktop wallpaper: if you display it on a 17" and a 19" LCD monitor, and both are set to a

35mm film is not suitable for poster size scans

If you have a 4000 spi scanner, you can, theoretically, use a 35mm negative to produce a file suitable for a 300 dpi 12" x 18" print. You should, however, not expect dazzling results. The extreme magnification will show any flaws in the original in extreme detail, and the scanner's own analog-digital conversion process will also increase image quality loss. 35mm originals are best for producing 4" x 6" or 5" x 7" prints. Medium or large format originals are better for producing genuine poster-sized results.

What's the Ideal Resolution for Printing?

Let's say you want to make a 5" x 7" print at 300 dpi. What image size in pixels do you need for the image file? Simply multiply the size in inches by the print resolution in dpi, and you will get 1500 × 2100 pixels. (Note that 300 dpi is the best relative resolution for printing with HP and Canon; for Epson it is 360 dpi.)

resolution of 1024 × 768, then you'll get one image displayed with a 17" diagonal and the other with a 19" diagonal. As we can see, the output size depends on the output device, or rather on its relative resolution. Depending on the chosen output device, the image requires different minimum relative resolutions in order to look sharp. Magazines, books, and prints from the lab use a print resolution of 300 dpi; inkjet printers use 200–300 dpi. Considerably lower values result in soft images.

Film Scanner Resolution and Maximum Print Size

Once you have bought a scanner, you'll want to make prints from the image files either with your own printer or with the help of a service provider. You'll want to know how far you can push the quality of your scanner. The size of the scanned file (in pixels) depends on its optical resolution (in spi, or as is more commonly used, in dpi).

From that you can calculate the maximum print size in inches. It is no problem technically to down-sample a large file, but it is not possible the other way around. If you have a 4000 dpi scan and scale it to 8000 dpi afterwards, you will not produce more picture information than was in the original scan. The only effect is that you waste a remarkable amount of storage space, as the 8000 dpi file is twice as large as the 4000 dpi file. Therefore, scanning is the key process that determines the quality of the image that can be achieved.

Always keep in mind, that the bottleneck in scanning resolution is in most cases the quality of the slide and not the ability of the scanner!

Sample Calculation in Three Simple Steps

You can calculate the maximum print size quite easily yourself as shown in the following three steps.

▸ First, consider the negative size. The bigger it is, the better. The nominal negative size of 24mm ×36mm is often a little smaller than 24mm × 36mm. For a Nikon F801s it is 35.95mm × 23.61mm. But it depends on the lens as well. To be on the safe side, the negative needs to be cropped a little to avoid black borders, since the scan preview is not completely accurate. Let's say we have now only 35.7mm × 23mm for scanning, which is 1.4055" × 0.9055" (1" = 25.4mm).

▸ Next, take the optical resolution of a Coolscan IV with 2900 dpi and multiply this by the cropped film size: this gives an image size of 4075 × 2627 pixels.

▸ To get the final print size, divide this image size by the print resolution of 300 dpi. 300 dpi has become the established print resolution standard for magazines and labs, so what is sufficient for this purpose should be sufficient for you as well. This gives a print size of 13½" × 8¾".

Crops Need High-Resolution Scanners

With a 4,000 dpi scanner you can print as large as 12.6" × 18.9" at 300 dpi. A 5400 dpi scanner lets you print up to 17" × 25.5" at 300 dpi. In practical use, an upgrade from a 2900 dpi film scanner to a 4000 dpi film scanner makes a big difference for large size printing – provided the scanners are otherwise comparable and the film stock is of very good quality. Much higher resolutions, such as 7200 dpi, make no sense since the film material cannot match that resolution.

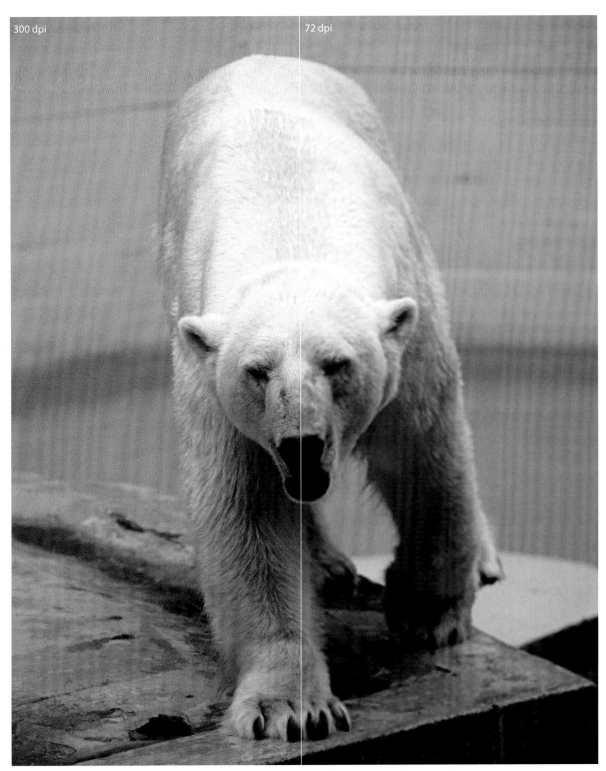

300 dpi

72 dpi

A 72 dpi resolution is sufficient for viewing images on a monitor, but you will need more pixels per inch for printing. A book or magazine image only appears really sharp when it is printed at a resolution of 300 dpi or more.

Improve Image Quality by Discarding Pixels

Here is a very simple trick to improve image quality: set the scan resolution to the optical resolution, e.g., 4000 dpi, and the scale to 50 percent (= 2000 dpi). This effect cannot be reproduced afterwards in an image editor; it has to be done during scanning. It is due to the effect that scanning in full resolution produces more noise, than scanning with a lower resolution. If you just scale down the scan afterwards, the noise is still there; at least in subjective matter, it gets even more visible through scaling. The best practice is to scan the slide at the same resolution you want to use later. If you scan primarily for archiving purposes, 2000 dpi is a good compromise.

Resolution of Scanned Image (24 × 36 mm)

Scanner Resolution	Scanned Image		Print Resolution	Print Size	Standard Lab Print
	Size in Pixels	Equivalent Megapixel of Digicam			
1700 spi	1 606 × 2 409	3.9 MP	300 dpi	5.35" × 8"	5" × 7"
			150 dpi	10.7" × 16"	
1800 spi	1 701 × 2 551	4.3 MP	300 dpi	5.67" × 8.5"	5" × 7"
			150 dpi	11.34" × 17"	
2900 spi	2 740 × 4 110	11.3 MP	300 dpi	9.13" × 13.7"	
			150 dpi	18.26" × 27.4"	
3200 spi	3 024 × 4 535	13.7 MP	300 dpi	10.08" × 15.12"	
			150 dpi	20.16" × 30.24"	20" × 30"
3600 spi	3 402 × 5 102	17.4 MP	300 dpi	11.34" × 17"	
			150 dpi	22.68" × 34"	20" × 36"
4000 spi	3 780 × 5 669	21.4 MP	300 dpi	12.6" × 18.9"	16" × 20"
			150 dpi	25.2" × 37.8"	20" × 36"
5400 spi	5 102 × 7 654	39.1 MP	300 dpi	17" × 25.5"	18" × 24"
			150 dpi	34" × 51"	
7200 spi	6 803 × 10 205	69.4 MP	300 dpi	22.68" × 34"	24" × 36"
			150 dpi	45.35" × 68"	

Color Management in Theory and Practice

An image passes through various devices on its way to final output. First the scanner reads the film into the scanner program. Then the scan is viewed on the monitor, processed in the image editor, and finally outputted on the printer. Each of these devices represents colors differently and has a different color gamut. Unintended color shifts are unavoidable without color management. This chapter explains what it takes to maintain consistent colors throughout the workflow.

5

Contents

5.1 Why Color Management?

A basic problem of digital image processing is that there is no absolute color definition across the entire workflow. In an ideal world, it would be possible to scan the color of the original accurately, display the scan on the monitor, and print it in true colors. Unfortunately, the current workflow cannot do that. The various links in the chain of digital processes represent colors in very different ways. This can give your image a color cast if you edit the image on an uncalibrated monitor and accidentally make modifications for what only seems to be a color cast.

Such problems can be avoided only with a solid color management system. Therefore, some basic knowledge of color models and color spaces is useful even for the photo enthusiast; for professionals, it is a must. This chapter will explore the topic of color management from the viewpoint of scanning film. Some theory will be introduced to help better understand the topic. The goal of the color management process is to interpret and reproduce colors consistently in every step of the workflow. An entirely correct representation of colors is not possible, since each device creates colors differently and therefore the color gamut of the devices used can visibly differ. It is a goal of color management to adjust the gamut as closely as possible.

5.2 Basics

The Relative Definition of Colors

In an 8-bit RGB image file, each pixel is described with the three basic colors; red, green, and blue, with 8 data bits for each. This allows for 2^8 = 256 different colors levels for a pixel in each of the three basic colors (2^{24} for RGB!). Thus, a red pixel can have 256 levels of red. A numerical value of 0 describes a complete absence of red, whereas the value 255 describes the fullest red that can be displayed. At this point a problem is already encountered with relative color definition: given two monitors, it is very unlikely that both will display the color red exactly the same way. For printing, it will be even more complicated. See yourself: take Photoshop and generate a uniform red area with the color values of red 255, green 0, and blue 0. Compare this area with the printed example on the opposite page. Due to the printing and the CMYK color separation, you will notice an obvious color difference, even though these patches have been generated with the exact numbers stated above in RGB mode.

Gamut

The range of colors that a particular device can reproduce is called its "gamut" or "color space." Theoretically, you can compare the color gamut of your scanner, monitor, and printer, and from the intersection generate a new color space in which each device would be able to represent every color. However, doing this would greatly underutilize the capabilities of each device. In the example mentioned above with the film scanner, monitor, and printer, the monitor displays fewer colors than the scanner can record. There are also some print colors that the monitor cannot display; but the monitor can also display many colors that the printer cannot handle. Using the smallest common color space would eliminate colors that could be scanned as well as printed. This is the reason that a device-independent color space is used as an interface between devices. However, at the various stages of the workflow, this color space has to be transformed several times. ICC profiles are used to keep the color representation consistent despite the transformations.

Gamma

Gamma controls the ratio of input to output signal in a display system (for example, a monitor). In the Windows world, a gamma of 2.2 has become standard; for Mac, it is 1.8. The gamma value controls how the mid-tones of an image file are displayed. The perceived brightness of a monitor depends on this value. For scanning you should use the standard gamma value of the operating system, which will be used later for editing and displaying the images.

Scanning, Viewing, Editing, and Printing

For a typical workflow, the first step is to scan the original with a film scanner. The resulting image file will be examined on a monitor. The monitor is the graphical interface for you, the operator. Based on the monitor image, you decide what corrections are needed. Once you are happy with the result, you can save the file with the changes and optionally print it.

Usually the colors in the print will not be identical to the colors on the monitor. This discrepancy is due to the different color models used by the various devices. The monitor works with the RGB color model, whereas the printer works with the CMYK color model. Inevitably, all these differences cause color deviations.

Unfortunately, even with color management these differences cannot be completely eliminated. After all, the monitor can display colors that the printer cannot generate, and vice versa. What color management can do, however, is match most colors, maintain a consistent color representation, and create at least a very good facsimile of the image.

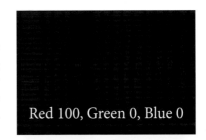

Red 100, Green 0, Blue 0

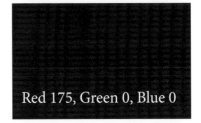

Red 175, Green 0, Blue 0

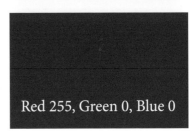

Red 255, Green 0, Blue 0

▲ In RGB mode, the red tones are defined with explicit color values.

5.3 Color Models

Color models are used to describe and classify colors. Color models also describe the method of generating colors. This method defines the palette of colors as a combination of a few primary colors. The RGB color model, for instance, generates all possible colors as combinations of the primary colors red, green, and blue. When all three color components take on maximum value, the result is the color white. With color models, it is not necessary to have a separate ink for every printable color tone. Actually, it would be impossible to provide a separate ink for every conceivable color tone. Instead, the entire color spectrum is covered by mixing a few basic colors. In this way, in combination with the basic color of the medium, all colors can be generated. For a monitor, the basic color is black; for printing, it is the white paper.

▲ RGB: The three primary colors – red, green and blue – add up to white.

Additive Color Model RGB

In the RGB color model, all colors are created from different amounts of the primary colors red, green, and blue. This model is called an "additive" color system. When all three colors are at their maximum, they produce white. Black is created by setting all three color components to zero; it is the absence of any light. Monitors and film scanners work with the RGB color model. A monitor that is turned off is therefore always black. For printing image files, the RGB color model has to be converted to CYMK. For professional printing, this happens in prepress. For printing on your home computer, this is taken care of by the printer driver. That means manual conversion to CYMK can usually be omitted. In fact, manual conversion should be avoided, since image information is lost that can be important for further processing. You should not convert images from RGB to CMYK without a good reason to do so.

Subtractive Color Model CMYK

The CMY color model consists of the three primary colors cyan, magenta, and yellow. It is a "subtractive" color model. The sum of these three colors yields black. If all color values are zero, no ink will be printed and the result will be white – assuming the printing paper is white. The problem with the CMY color model is that a proper black can be obtained only from absolutely pure primary colors: cyan, magenta, and yellow. Even though three expensive layers of ink are put down, one usually gets only a dirty black-brown instead of pure black. It is considerably less expensive and better to add black as a spot color to the primary colors. This combination is called the CMYK color model (the "K" stands for "Key"). In this fashion, a single layer of ink can generate a deep black.

▲ In the CMYK color model the sum of the three primary – colors cyan, magenta, and yellow – is black.

Image: Almute Kraus

Lab

The Lab color space contains all colors that can be seen by the human eye. The three values L, a, and b define the color in a precise way. "L" stands for luminance, "a" stands for the red-green component, and "b" stands for the blue-yellow component – which is not very intuitive for most people. The color spectrum of the Lab model contains all colors of the RGB and the CMYK color spaces. Lab can therefore define colors precisely.

Unlike RGB and CMYK, the Lab color space is device-independent. However, it has to be converted first to either RGB or CMYK in order to be displayed or printed on any device. Without an output device such as a monitor or a printer, the user cannot do much with abstract color information. One practical use of the Lab color model is for describing color deviations precisely; for example, when calibrating devices. Also, the Lab color space is used as a transfer color space for converting images from one color space to the other. In that case, image files are first converted from the original color space to Lab, then from there to the target color space.

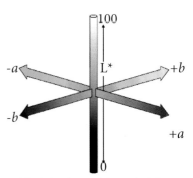

▲ The Lab color model exactly defines colors mathematically.

Up-to-date inkjet printers still use cyan, magenta, yellow and black ink. Moreover some of them add extra ink tanks like light cyan and light magenta (e.g. Epson Stylus Photo 1400) for enhanced brillance.

Image: Epson

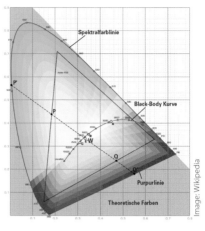

The CIE XYZ color system, illustrated using the so-called "shoe sole" diagram, represents all colors visible to the human eye. Individual color spaces are represented by subsections of the space within the shoe sole:

5.4 Color Spaces

No device can display all colors visible to the human eye. That is why there are color spaces; they represent a limited section of a color model. For example, the sRGB color space is just a subset of the RGB color model. A color model can cover several color spaces. Therefore, there is not just one color space, but several color spaces per color model, each with a specific purpose. You should choose a color space just large enough to fit within the color space of the device. If the color space substantially exceeds the color range of the device, the bit values for non-displayable colors will be wasted. This means that you will give up intermediate tones in the range that can actually be reproduced.

Considerations for Selecting a Color Space

You can choose among many color spaces. However, there is no ideal solution; it will always be a compromise. Color spaces differ in the color gamut they can reproduce: the larger the color space, the larger the number of displayable colors and the more flexibility for editing. However, not all devices have a large gamut. Therefore, before output, images have to be adjusted to the gamut of the output device. If you use a color space with a smaller gamut, such as sRGB (optimized for monitors), you will not have this problem. The colors can be reproduced on most output devices without problems, and the colors appear more vivid. You pay for the adjustment, however, by giving up some colors. Moreover everything depends on your specific equipment. There are some high-priced TFT screens around that can fully display the Adobe RGB color space. You can see every color covered by the Adobe RGB on your screen, which is quite handy when you do image editing. An affordable monitor will surely lack this specific ability.

sRGB, Adobe RGB, etc.: Color Spaces for Gamma 2.2

The sRGB color space is the general standard color space for display monitors on Windows PCs. It does not describe the gamut of a particular monitor, but rather an average for most monitors. Originally, it was intended as the standard color space for the Internet.

At first sight, it looks like an ideal solution for digital image processing, but there are disadvantages: the scanned original can contain colors that are not included in the sRGB color space. If these colors are within the gamut of the scanner and the printer, then using the sRGB color space will waste them just because they cannot be displayed on an average monitor.

sRGB is ideal for users who do not do post-processing and who only display their images on a monitor. The disadvantage of sRGB is its small gamut. In comparison to other color spaces, this can have a negative effect for printing.

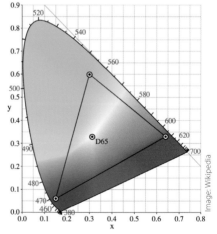

The sRGB color space represents only a very small portion of the colors described by the shoe sole diagram, and is thus ideal for use on the Internet.

The following are just a few of many other color spaces in the RGB color model:

▸ sRGB
▸ Bruce RGB
▸ NTSC (1953)
▸ Adobe RGB (1998)
▸ CIE RGB
▸ Wide Gamut RGB
▸ Wide Gamut RGB (compensated)

The above color spaces are listed in order of progressively larger gamut: sRGB has the smallest and Wide Gamut RGB (compensated) the largest. Apart from sRGB, Adobe RGB is also very widely used by photographers. It has a distinctly larger gamut than sRGB and offers more possibilities for editing and printing.

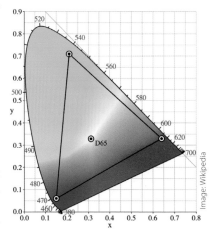

▲ Adobe RGB is capable of reproducing many more colors than the relatively limited sRGB color space.

Apple RGB, etc.: Color Spaces for Gamma 1.8

In the Apple world, everything is a little different from the Windows world. This also applies to color spaces. Apple color spaces use a standard gamma of 1.8. The following list is sorted by gamut:

▸ Apple RGB
▸ Color Match RGB
▸ Apple RGB (compensated)

Apple RGB has the smallest gamut and Apple RGB (compensated) has the largest gamut.

Conversion between Color Spaces

Since every device has a different color space, the color space has to be converted several times in the workflow. The RGB color information generated in a scan depends on the gamut of the scanner. This device-dependent color space should be converted to a standardized working color space, such as sRGB. The working color space is used for editing the image, provided that the image editor is configured accordingly. In order to print the image, another conversion to CMYK is needed. On a home computer, the printer driver takes care of this conversion without any user intervention. In every case, there will be several conversions between color spaces in the course of the workflow. The conversions go from device specific to device independent, and vice versa. By using color management systems with ICC profiles, the results of the conversions will not be arbitrary, and they can be controlled.

image in device-independent CIE color space

internal color space

image file (RGB)

Image: Jürgen Gulbins

▲ The color management module of the color management system converts the color spaces and provides consistent colors among different devices.

5.5 ICC Profiles

Purpose of ICC Profiles

The International Color Consortium (ICC) has worked on the problem of how to transform images between different color spaces with the least possible quality loss. The result is a standard based on so-called ICC profiles. As already mentioned, these profiles control the transformations between color spaces. There are three different types of profiles:
- Input profiles
- Output profiles
- Display profiles

Input profiles transform the scanned color information – which at this point still depends on the scanner – into a device-independent color space. The transformation does not require ICC profiles; these are optional. ICC profiles are used to compensate for device-specific deviations. The measured deviations from standard values can be ironed out with their help. Devices are calibrated to standard values. With calibration, one can ensure that a particular color value will be recognized by the device as such.

Definition of Calibration and Profiling

The terms "calibration" and "profiling" always come up in discussions about color management. Often they are used synonymously, although they are actually two different processes. For scanning, calibration means simply measuring the deviations against a certain standard. Usually an IT8 calibration target is used to accomplish that.

Knowing that the scanner has some color deviations does not do much good by itself, but the data obtained by the calibration process can be used to profile the device. With a device profile, any deviations between measured values and standard values can be corrected. For scanners, ICC profiles are used. Since in common usage the terms "calibration" and "profiling" are not clearly separated, I will use them interchangeably in this book.

Input Profiles

Input profiles are used for input devices, such as scanners. A scanner input profile enables color values extracted from the scanner to be translated to device-independent CIE Lab color values. The scanner is usually delivered with some color shifts due to normal production tolerances. Generally, it is not possible to scan the film with a high level of color accuracy. The scanner will recognize red as red, but it is rather unlikely that a precisely defined standard value will be recognized as exactly that value. In practice, small errors are unavoidable.

To get exact readings of the deviations, one can scan a so-called IT8 target, which is a tightly controlled color chart on either transparent or reflective stock. The image data from the scan can then be compared to the reference file supplied with the target. Most likely there will be some deviations, and these will be captured in the ICC profile. If a scanning program that supports ICC profiles is used, these deviations will be automatically corrected. True-color scanning requires individual profiling of the scanner.

The International Color Consortium (ICC) can be found at www.color. org. The specifications of all standards published by the ICC can be downloaded from the website.

▲ Applying a scanner and monitor
profile in VueScan.

Output Profiles

Output profiles for the output device are the counterparts of input profiles for the scanner. Output profiles enable colors from a device-independent color space to be translated to the color space of an output device, such as a particular printer.

Apart from the printer itself, the key factors for the color characteristic of a print are the ink and paper consumables used as well as the settings in the printer driver for print resolution and other parameters. An ICC profile for a printer is valid for only that specific combination of parameters. For profiling, a reference file is printed – a printer target – and the printed colors are measured with a spectrophotometer. The ICC printer profile is calculated from the differences between the nominal values of the file and the measured values from the print.

This profile adjusts the differences encountered when an image file from the device-independent color space is converted to the device-specific color space. Without an ICC printer profile, accurate colors cannot be printed.

Display Profiles

During editing, the monitor is the output device. To be displayed on the monitor, the color information of the image file has to be translated to the device-specific color space. The monitor gives important feedback to the user during editing and can even act as an input device when coupled with a touch screen. The image can be checked and modified, and then the result saved. After editing, the device-dependent color space has to be transformed back into a device-independent color space. For this reason, display profiles are also called internal profiles.

Color Management in Operating Systems and Application

Microsoft and Apple have each integrated a color management module into their operating systems. Microsoft calls it ICM; Apple calls it ColorSync. The role of this module is to link the device profiles. Integrating the module into the operating system allows for speedier processing of processor-intensive conversions.

Applications can also use their own color management module. The Adobe Color Engine (ACE) is one example of a module tailored specifically to Adobe applications. Other applications normally use the resident operating system's module.

◀ The monitor profiles can be assigned with Windows utilities.

Download ICC Profiles or DIY?

Many manufacturers supply ICC profiles for download or include them on the CD packaged with the device. So, do you actually have to create your own ICC profile? After all, it takes time and money to do so. The advantage of using the manufacturer's ICC profile is obvious: you simply download and install it. However, the quality of these profiles varies quite a bit.

Devices change their color characteristics as they get older. For that reason, you should refresh your self-made profiles regularly. Apart from aging, there are also manufacturing tolerances that the manufacturer's profiles cannot cover. A manufacturer's ICC profile is better than no profile at all, but you will only get optimum color accuracy with an individually-generated profile, which must be regularly updated.

▲ QuickGamma is a small tool for software calibration. You can use it together with QuickMonitorProfile of the same manufacturer for basic monitor calibration free of charge

5.6 Monitor Calibration in Practice

Without a calibrated monitor, image editing is rather difficult or almost impossible. It is difficult to judge whether you are trying to correct a flaw in the image file or whether you are being misled by an erroneous display on the monitor. There are various hardware and software tools for monitor calibration. Hardware-based tools consist of a measuring device and special software. In the case of software tools, the human eye serves as the measuring device.

Truly accurate results are only possible with a hardware tool. Such a tool measures screen colors, compares them with nominal values, and calculates the ICC profile from the differences. For consistent quality, calibration and profiling should be repeated monthly. Software solutions rely on your eye.

Software calibration methods are far less accurate than hardware calibration methods, since the color adaptation of eye and brain depends on the ambient lighting conditions. Color and contrast adjustments are done directly on the monitor's controls without generating the ICC profile. If you do not want to spend money for a hardware-based calibration tool, with software you can at least improve some settings.

Monitor Calibration via Software

There are numerous calibration tools on the market. None of them are as good as hardware calibration, but all of them are better than no calibration at all. QuickGamma is a well known freeware application, but it is very basic and does not really adjust colors, only their gamma. You can download it at http://quickgamma.de. There you will find another freeware program that is called QuickMonitorProfile. You can use it to create monitor profiles, it supplements QuickGamma.

DisplayMate is a more sophisticated program, but it is not freeware. The website www.displaymate.com no longer offers a demo download. This tool costs $69 – it is strong on diagnostics but weak on calibration. Before you pay this, check out Calibration Wizard at www.hex2bit.com. It is free, although a commercial, more sophisticated version of Calibration is planned for the future.

▲ Calibration Wizard is freeware, but it will allow you only to adjust gamma brightness and contrast.

Applications & Color Spaces

No matter what picture application you use, whether an editor or just a viewer, in most cases it will have its own color space, too. You can configure it in the properties section. All applications you use should be configured the same way; otherwise, you will have differences in color reproduction on the screen despite calibration.

Monitor Calibration via Hardware with ColorVision Spyder 2

ColorVision offers solutions running under Windows and Mac OS that work for CRT monitors as well as for LCD monitors. They range from the ColorVision Spyder 2express to the more expensive Spyder 2PRO, which comes with additional adjustment options. Each version contains the same colorimeter but uses different software. The colorimeter plugs into a USB socket, which is where it also draws its power. The basic calibration functions are the same for both versions. The software generates test patterns on the screen, and the colorimeter takes readings off them. The software then generates the ICC profile based on these measurements. Once this profile is assigned to the display card, the monitor is successfully calibrated.

ColorVision provided me with a Spyder 2PRO for testing. The following section refers to the Spyder 2PRO for calibrating a CRT monitor. The less expensive Spyder 2PRO model from ColorVision offers fewer adjustment options than the tested version. You should first install the software before plugging the colorimeter into the USB port of your PC, and reset any of the changes you may have made to the display card driver's default settings. Also, the monitor should be reset to its original factory setting via the on-screen display. To use the colorimeter on CRTs, you need to remove the LCD baffle (see image below). For LCD monitors you leave this baffle on. Software installation is simple; just make sure the colorimeter is not plugged in when the software is installed. After restarting the computer, you can plug the colorimeter into

Image: ColorVision

▶ The Spyder 2 colorimeter can also be used with CRT monitors.

High-Bright, SuperBright, UltraBright, etc.

Some CRT monitors offer several modes to optimize the display for best results with either text, video, or images. These modes affect the White Luminance. Pick one mode and stay with it when calibrating the monitor.

the USB port and start the software. The software offers detailed help for each adjustment. Once you select monitor type, you are prompted for the desired gamma and white point. The default calibration is set for a gamma of 2.2 and a white point with a color temperature of 6500 Kelvin. Spyder 2PRO also offers the possibility of setting luminance values freely, although it is not necessary for adjusting the Visual Luminance.

Next, adjust the White Luminance of the monitor. This is done in hardware with the monitor's contrast control, which varies from model to model. In most cases, it is done with the on-screen display (OSD) of the monitor (consult the monitor's user manual for that). Under no circumstances should you use the display driver to adjust White Luminance, even if there is a provision to do so. The contrast needs to be set so that the four patches of the test image can still be separated.

Next, adjust the Black Luminance with the monitor's brightness control. This is also done via the monitor OSD. The adjustment is perfect if you can barely distinguish between the four patches of the black test

Avoiding Software Conflicts
To avoid software conflicts, disable other manufacturers' software for calibrating monitors, such as Adobe Gamma. If you have installed Photoshop, you should disable it in the Startup tab of Windows System Configuration Utility (run msconfig).

Image: ColorVision

◀ For CRT monitors, you need to remove the LCD baffle.

Selecting ColorVision Monitor Profiles under Windows

You can select the monitor profiles either with the ProfileChooser supplied by ColorVision or directly in Windows. For the latter, right-click on a blank area of the XP desktop and select Properties > Settings > Advanced > ColorManagement. In Windows Vista, right click a blank area of the desktop and select Personalize > Display Settings > Advanced Settings > Color Management > Color Management. It is not really necessary to install the Profile-Chooser. In either case, the ICC profiles will be available to all Windows applications.

image. Now the actual measurements can begin. Place the colorimeter on the screen - the software generates a sequence of test patterns in different colors for the colorimeter to read. The measured values are compared with the nominal color values and an ICC monitor profile is generated. You can name this profile whatever you like. It will be saved as an ICM file under C:\WINDOWS\system32\spool\drivers\color\. You may have to modify that pathname if your installation path for Windows is different.

Once the monitor is properly calibrated, you can assume that your images will be displayed correctly on the device. The same applies to any other monitors calibrated with the same gamma and color temperature. However, it does not mean that images optimized on your monitor will be printed with identical colors by your inkjet or your photo lab. Monitor calibration affects only the image display on your monitor and helps you produce correct colors and tones only during editing.

▶ Hardware monitor calibration is standard for professional users, despite the cost and effort it takes.

Image: Quato

◀ ▲ Proper positioning of the mask is crucial for accurate IT8 profiling.

5.7 Generating ICC Scanner Profiles

To generate an ICC scanner profile, you need a scanner target and appropriate software. For film scanners, the target is a calibrated color slide – an IT8 target. The scanner turns the color chart into a normal scan file. The profiling software compares the scanned color values with the nominal values from a reference file and calculates the profile. Scanning software, which can use ICC profiles, can create scans with high color fidelity.

Scanner Calibration with SilverFast

The full version of SilverFast Ai from LaserSoft Imaging includes an IT8 calibration slide. This is a convenient solution, that VueScan can not match. But of course SilverFast is not exactly what you would call a cheap software. Generating a profile is conveniently integrated into the software and only takes a few mouse clicks.

First, insert the slide into the scanner and generate a preview. Next, draw a selection frame. This must be placed precisely or else the measurements will not be accurate. If the mask position is off, the program will display a warning. Even a small misalignment can match color patches with the wrong reference values and thereby corrupt the profile. If a measuring field overlaps with more than one color patch, a faulty profile will be generated. Small overlaps may not be noticed automatically by the program.

After measurements are compared with reference data, a profile is generated and saved in Windows XP or Windows Vista under \WINNT\ system32\spool\drivers\color.

▶ Nikon Scan uses built-in scanner profiles. As a result, this image has a slight color cast.

▶ It is better to generate an individual profile with an IT8 target. With VueScan the same image shows more neutral color reproduction.

▶ In both cases, it still takes some editing for a satisfactory result. Calibration will not save you this step.

Scanning Methods

Scanning is done in several steps. First, a thumbnail index of the filmstrip is generated. From this index, a preselection is made. In the next step, previews of the selected images are generated. Then, the preview must be edited to prepare the image for the scan. This chapter describes these steps in detail. There are hardly any differences among Nikon Scan, VueScan, and SilverFast in the scanning procedure. For best results, none of these steps should be skipped; but for high-volume processing, a step might be skipped to save time.

Contents

▲ SilverFast Ai does not let you rotate the thumbnails in the Overview, but Nikon Scan will.

6.1 Thumbnail Index – Filmstrip Offset and Presorting

The index scan is the quickest way to get an overview of the inserted filmstrip. The thumbnails on the index are low-resolution and can only give a rough idea of image quality, but they are good enough to presort and to correct any filmstrip offset.

For negatives with normal exposure, there is a pronounced contrast between image and frame gap. For these, automatic image positioning works reliably. However, it is a different story for night shots, where image background and frame gap have almost the same density. In those cases, the frames may have to be manually aligned. Many programs from scanner manufacturers do not provide for manual alignment, but VueScan, SilverFast, and Nikon Scan are all capable.

The most convenient place to rotate or mirror the images is in the thumbnail view, although it is still possible later in the preview or final scan. Rotating in the thumbnail view prevents the possibility of creating reversed images by mistake.

The thumbnail index is only available if there is more than one image in the scanner. This feature simplifies batch processing. VueScan cannot generate thumbnails but goes straight to the preview instead. However, it can save a thumbnail index in parallel with previews and scans. Silver-Fast can save thumbnails directly, but Nikon Scan cannot.

6.2 Preview – Corrections Before the Scan

Preview (also called prescan in some applications) takes more time than the index scan. The preview for a single image can take longer than the index scan of a whole filmstrip, but you get a much larger preview image as a result. It lets you edit the image enough to give you the best quality scan afterwards.

It makes sense to trim the image in the preview window. By default the scanner captures the entire image plus border and frame gap. If you restrict yourself to the actual image area, you will get more compact

◀ The preview is much more suitable for evaluating the image than the tiny thumbnail.

image files. In addition, in Nikon Scan the black borders interfere with auto exposure. SilverFast and VueScan both have an elegant solution for excluding the borders for exposure metering.

Cropping the Image

For scanning, only frame edges should be trimmed. At this stage, the image should not be cropped; rather, the entire image information of the original should be preserved.

Later, during editing, the image can be cropped for aesthetic improvements. If the image is cropped, it may have to be rescanned later. Images should not be cropped, especially if they are being scanned for archival purposes.

I do not recommend that you leave the cropping frame at its (default) full-frame setting. Doing this includes the black borders of slides and negatives in the scan, and can lead to incorrect exposure. The scanner will regards the black borders as part of the image and adjust its exposure accordingly. This will result in flatter colors and generally less vivid results. Due to this, it is better to set the cropping manually before you start scanning.

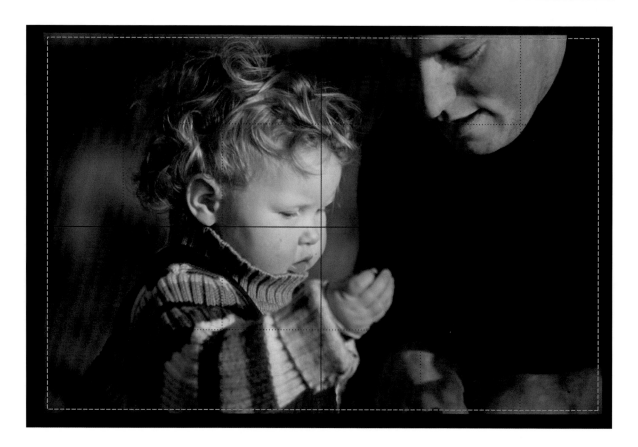

▲ If possible, you should trim the black borders before the scan and not after.

Image Corrections Before the Scan

Image corrections such as ICE, FARE, ROC, or GEM can be applied in the preview so that their effects can be assessed in the preview window. The extent of the corrections depends on the chosen operating mode. The following adjustments must be performed before the scan with all scanning programs, since these corrections cannot be applied afterwards:

▸ Analog gain / exposure time
▸ Focusing (AF or manual)

If images are scanned in TIFF format, all the corrections will be permanently included in the image file. RAW files are more tolerant in that respect, but even there some adjustments are irreversible. For instance, in Nikon Scan the following corrections will be permanently included, even with NEF:

▸ Dust and scratch removal with Digital ICE
▸ Color restoration with ROC
▸ Highlight and shadow recovery with DEE
▸ Grain reduction with GEM

If any of these corrections are unsatisfactory, the image will have to be rescanned.

Difference Between Scanning Program and Image Editor

Certain adjustments are possible with the scanning software, but in most cases they are not useful. Unsharp masking is one example. This should be done only if the resolution of the final image is already fixed. Usually the image resolution is not identical with the scan resolution.

If you work with Nikon's RAW format NEF, you will not have to commit yourself immediately. Choices are only saved as additional information in the scan file and can be changed later as often as desired. Unlike with TIFF and JPEG files, the configurations are not permanently applied to the image data. Therefore, at RAW level, the risk of degrading the image data with poor adjustments is lower. Don't spend too much time with fine-tuning color and contrast in the scanning software; doing so just disrupts your work rhythm. Only apply essential corrections and use the image editor for optimizing the image during post-processing. Best practise would be to apply only adjustments during scanning that you have to, like Analog Gain. Most corrections can be done more conveniently after scanning.

▲ NEF files let you freely change the unsharp masking afterwards.

Sorting Your Slides with Loupe and Lightbox

It is much quicker to sort out unsharp slides before you insert them into the scanner. A high-quality 8x magnifier, such as the ones offered by Rodenstock, Schneider, or Pentax, is good for that. If possible, use a lightbox as well. Examine your slides carefully and you will not waste time with unnecessary scans.

Scan – Create the Image File

After preparations with index scan and preview are done, the actual scan is comparatively simple. All that is left is to pick a meaningful name for the image file and select the proper file format. I always try to name the files with numbers that correspond with the filmstrip numbers. This makes it easier to clearly match file and original. In addition to naming your files, you have to choose a file format, which should be either TIFF or RAW.

Maximum Color Depth for Maximum Quality

At this point, you have one last chance to change the color depth. Often, scanning programs are set to a default of 8 bits per color channel. However, you should take advantage of the maximum color depth supported by the scanner's A/D converter, which is usually 12, 14, or even 16 bits. Afterwards, you still can reduce it to a lower bit rate.

The scanner software lets you define the final file name. Apart from progressive numbering, you can also define prefixes and suffixes for the file names. If you don't have a perfect numbering scheme, don't waste too much time - with freeware tools such as Lupas Rename, you can quickly and easily rename large numbers of files later.

* Lupas Rename is available free at http://rename.lupasfreeware.org. The program contains much more powerful file renaming tools than those included as standard with Windows operating systems.

Single and Batch Scan

Going through index scan, preview, and scan can take several minutes. Most of the time is spent not on setting up each scan, but on the scanning process itself. Depending on the model, this can be a real test of patience. It is better to prepare several images at once and have the scanner run in batch mode unattended.

Only in batch mode can you manage to scan large numbers of slides or negatives. Some scanners support batch scanning in their basic configuration. For other models, optional adapters are available; some models cannot do batch scanning at all. Typically, the supplied slide holders can batch four to six slides. Even more convenient is scanning directly from slide trays, like the Reflecta DigiDia 4000 (identical to Braun).

Don't Frame Your Slides...

Nikon 35mm scanners are rather rudimentary in handling mounted slides; in their basic version, they can only scan one at a time. You will need the expensive slide feeder attachment SF-210 for the model Super Coolscan 5000 ED to batch-scan mounted slides. Competitors such as Minolta and Reflecta offer more user-friendly solutions. For Nikon, the least expensive way to batch-scan slides is to leave them uncut in the strip, since the supplied filmstrip adapter SA-21 handles color positives as well. This makes it easier to archive the slides, too.

6.3 Multi-Sampling/Multi-Exposure

To improve image quality, a slide can be scanned multiple times. The scanner software averages multiple readings. Multi-sampling is used to reduce image noise, to get more natural colors, and to achieve smoother color transitions.

There is a difference between single-pass and multi-pass multi-sampling. The Coolscan 5000 ED uses the single-pass method; each line is read multiple times before the scanner advances to the next step. In the multi-pass method, the entire image is first scanned completely and then repeated multiple times. With the multiple passes the scanning time increases accordingly. Coolscan IV/V do not support multi-sampling under Nikon Scan. With SilverFast and VueScan, both models can do multi-sampling, but only in multi-pass mode. It is a functionality that Nikon did not intend to deliver with cheaper scanner models. As Coolscan IV/V share the same hardware as their expensive brothers Coolscan 4000/5000, the absence of single pass multi-sampling has no technical basis; it is an artificial restriction of the manufacturer.

* If you want to use multi-sample scans, your original must be firmly fixed in position in order to avoid slippage during the multiple scan sequences. Auto-loading strip film adapters are not suitable for this type of application, and it is better to use a stationary film holder or to individually frame your original.

Images: Minolta

◀ Minolta DIMAGE 5400: on the left the insertion of a single slide; on the right the supplied slide holder SHM10 for up to four mounted slides, which makes for a rather small batch scan.

Image: Peter Steinhoff, ca. 1977

◀ ◀ With a single scan, the texture of the film is quite visible.

◀ With multi-sample scanning (16x), the transitions become smoother, but the scanning time is several times longer than for a single scan.

Multi-sampling is used only for squeezing the last bit of quality from an original–it comes in handy, for example, when scanning old slides. While multi-sample scans expose each scan identically, multiple exposures mix a number of scans with differing exposures to achieve the final result. This technique can be used to visibly increase a scanner's dynamic range; and has a more pronounced effect in cheaper scanners than in more expensive models that have already a bigger dynamic range. VueScan is capable of multiple exposures, as are the more expensive SilverFast Studio and SilverFast SE Plus packages (version 6.5 and higher). Nikon Scan is not capable of making multiple exposures at all. It only supports multi-sampling for Coolscan 5000 and Coolscan 9000.

▼ If the scanner is slightly out of registration in the multi-pass mode, it will cause banding, as shown in this image. The more you down-sample the image, the more obvious the effect will be.

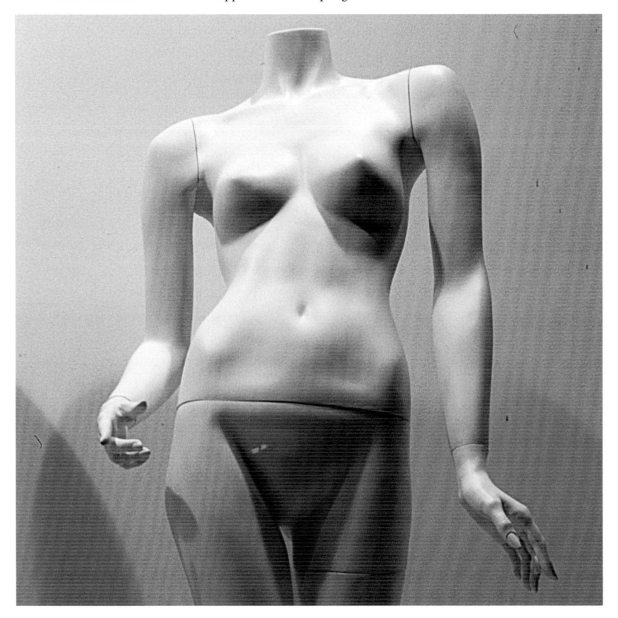

Scanning Correction Filters in Detail

Modern scanners can read the tiniest bit of information on an image. Unfortunately, they also record details we could do without: specks of dust, scratches, film grain, and the color flaws of faded slides and negatives. The first time I saw my slides on a monitor, I was shocked to find how many scratches there were. No matter how carefully you handle it, every piece of film has some minor or major damage. Powerful scratch removal functions in hardware and filters for smoothing film grain and restoring faded colors are standard nowadays. However, there are big differences in the way Nikon Scan, SilverFast, and VueScan handle these functions. In almost all other manufacturers' scanning programs, image correction filters are poorly implemented.

7

Contents

7.1 Dust and Scratch Removal in Software

*Scanner Light Sources and
Scratches*
*The scanner's light source plays a
major role in highlighting scratches
in black and white negatives. The
comparatively harsh LED light of
Nikon scanners tends to emphasize the scratches. This effect is less
pronounced in scanners with soft,
diffuse light sources. Using the right
scanner model can reduce the need
for scratch removal.*

After scanning, there is always an opportunity to touch up dust and scratches later in Photoshop. Most image editors can handle these functions, and for Photoshop there are also convenient plug-ins. However, software cannot easily distinguish dust and scratches from other image details. Therefore, the process cannot be automated. In some cases, manual retouching makes sense; it gives full control over the extent of correction required to improve image quality. For large numbers of images, however, it is too time-consuming and therefore not practical. Scratch removal in software is used only when hardware-based methods fail.

SilverFast SRD

The software program SilverFast Ai offers its own scratch removal with SRD. During the scan, it analyzes the entire image and automatically marks the scratches in red. This method is highly recommended for black and white film. For color film, SRD makes sense only for scanners without ICE, which are rarely recommended anyway.

SRD is clearly inferior to ICE in almost every regard. The tedious process of manually marking scratches can only partly be avoided with the automatic detection feature. The sliders Detection, Defect Size, and Intensity let you configure automatic detection. In the default setting, SRD will not detect the infamous horizontal "telephone wire" scratches on the film surface. For that problem, there is an expert mode with different settings that increase the detection rate of horizontal or vertical scratches quite a bit.

In my tests, I generally got the best results with the correction set stronger. Some detail gets lost during the process, but the annoying scratches are mostly gone. Overall, SRD simplifies scratch removal in some ways, but the detection rate could be improved. Even with automatic detection, there is still always much need for manual retouching. Due to this, SRD is only an alternative when ICE does not work, as is the case with most black and white negatives. If you are an experienced Photoshop user, it can be easier to correct dust and scratches manually than to use SRD in SilverFast.

A long horizontal scratch and specs of dust spoil the black and white negative.

The horizontal scratch is detected only when options are set properly.

▲ SRD and iSRD let you protect delicate areas with masks. In the case of SRD, the software incorrectly detects the catch light on the lips as a scratch if you do not cover it with a mask.

SilverFast iSRD

LaserSoft Imaging has a scratch removal method that uses infrared for detection but still allows for a large degree of manual intervention. Quato has implemented this method in its IntelliScan 5000 (the images printed on the opposite page are from this scanner). While ICE and Infrared Clean offer little control, iSRD can be configured more flexibly. The familiar SRD still exists in parallel and even has a menu in the same window, but the two methods cannot be used simultaneously.

In actual use, this supposedly great masking method quickly shows its limitations. This is because it subdivides the preview image into many small segments. The program takes its time going from one segment to another, and many seconds can elapse before the image is ready to be processed. For proper masking, each segment must be inspected; it is quite easy to spend 10 minutes on checking just one image alone. Also, the sliders help improve the detection rate of iSRD a little bit, but in my test negative, there were still many scratches that even the highest setting in iSRD missed.

iSRD can be run in automatic mode as well as in the manual mode described above. Using automatic mode helps avoid having to set a lot of troublesome configurations; but overall, this new correction scheme cannot compete with ICE and Infrared Clean. The idea is good, but the execution – at least with the Quato 5000 scanner – is poor. According to SilverFast, this is a hardware based problem of this specific configuration, as the Quato scanner (manufactured by Plustek) is supposed to have an imprecise infrared channel. It should work better with a more sophisticated scanner like a Nikon Coolscan 5000, though.

◀ iSRD lets you fine-tune the correction filter with sliders.

The uncorrected image is littered with scratches; it urgently needs dust and scratch removal.

The iSRD scratch removal of the Quato 5000 scanner is not as thorough as ICE.

7.2 Dust and Scratch Removal in Hardware

Digital ICE

Software-based correction is not as powerful and user-friendly as systems built into the scanner hardware, whereby an infrared beam detects dust and scratches to be removed from the image data. Scanner hardware and software have to support the inherent correction method together. For that reason, there is no ICE plug-in for Photoshop. Digital ICE successfully removes dust and scratches in Nikon and Minolta scanners. The process works only with color negatives and color slides. For a Nikon scanner, one of the following must be selected before the scan:

‣ No ICE
‣ ICE normal
‣ ICE fine

ICE is CPU-intensive, so scanning times and processor load increase considerably. Despite that, ICE should always be enabled. One disadvantage of using ICE is some loss of sharpness and detail, especially in ICE fine mode for removing even the finest scratches. ICE normal mode is sufficient for most cases; almost all scratches will be eliminated without too much loss of sharpness.

The softening of the image is justifiable and is fully recoverable with sharpening. Not every product that is labeled with ICE also contains the full ICE. There are flatbed scanners from Microtech that offer ICE only for reflective media, but not for film where it is needed mostly.

Silver Crystals Block Infrared Detection

Conventional black and white film contains silver crystals, which prevent dust detection via infrared. These films are explicitly not supported by ICE. This does not apply to chromogenic black and white films that can be developed with the C41 process for color negatives. Kodachrome slides are also special; they are based on a different technology (K14) than normal slide film. The standard versions of ICE should be turned off for Kodachrome slides since they degrade image quality. Currently, only Nikon's Super Coolscan 9000 ED comes with Digital ICE professional, which can handle Kodachrome.

With a Scanhancer, even regular ICE works for Kodachrome. The following image is an extreme case; the negative has many deep scratches. The occasional scratches and specks of dust that are present, even in carefully handled film, are completely removed by ICE. By default, ICE should be set to normal for standard scans. Only in exceptional cases should it be turned off.

ICE fine is only rarely useful. It does remove the finest scratches, but the image suffers noticeably. Alternatively, Photoshop may be used to remove any scratches that ICE normal misses. Compare the detail views of the scanned images on the right.

◀ Without ICE: serious scratches ruin the image almost completely.

◀ With ICE normal: most scratches are gone. Unsharp masking can compensate for the slight loss of sharpness from the scratch removal.

◀ With ICE fine: The finest scratches have been removed, but the image looks soft. ICE fine could not remove the deep scratches completely, though. Therefore, this type of heavy correction is in general not recommended.

Top row: The enlarged detail view shows the increasing effect ICE has on image sharpness, although at normal magnifications it does not look so bad.

Lower row: Infrared Clean from VueScan is not as effective as Digital ICE, but it preserves the texture of the image even at the strongest filter setting.

ICE and FARE for Kodachrome Slides

Kodachrome film, based on the K14 process, is a different technology than the conventional E6 slide film. The emulsion side of Kodachrome has a similar appearance to black and white film. As already noted, ICE does not work for black and white film, so it comes as no surprise that scanner manufacturers also advise against using ICE on Kodachrome.

So far, the only scanner that offers ICE specifically for Kodachrome is the Nikon Super Coolscan 9000 ED ($1,900). I have not tested it myself, but this multi-format scanner has generally received positive reviews.

For my tests on Kodachrome slides, I had to make do with a Nikon Coolscan IV and Coolscan 5000 scanner. In my first attempts at scanning Kodachrome, ICE was accidentally turned on. In full-screen mode the result actually looked acceptable. Unlike with black and white negatives, with Kodachrome film, ICE actually removes dust and scratches. The subjective quality of the correction was not overwhelming, though. This type of correction works better with E6 and C41 film.

A further problem is that ICE can mistake fine image details for dust and remove them. This becomes visible only when you zoom in. Tests reported by a German computer magazine showed that this effect varies with the scanner model: the ICE of the Minolta DIMAGE Scan Elite 5400 could keep up with the ICE professional of the Nikon Super Coolscan 9000 ED when correcting Kodachrome, even though the Nikon is optimized especially for Kodachrome. So, ICE and FARE should not be automatically dismissed for Kodachrome.

The best recommendation is to simply run your own tests to see the effect of ICE. Also, if you are not planning on big enlargements, it makes sense to sacrifice some minor details for the sake of a generally clean image. (Note: With a Scanhancer, regular ICE works for Kodachrome without limitations.)

Top row: Without ICE, annoying dust on Kodak T-Max is apparent.

Lower row: With ICE applied, the dust is still there, but the texture of the iris is completely gone.

ICE and FARE for Black and White Negatives

For physical reasons, scratch and dust removal does not work with conventional black and white film like Kodak T-Max. The silver crystals in the emulsion are identified incorrectly as dust particles, which leads to faulty correction. Jus take care that ICE is disabled when scanning black and white film to avoid massive loss of details.

Top row: Without ICE; the beluga whale on Ilford XP2 (C41 process) film shows clearly visible dust and scratches.

Lower row: With ICE; scratches and dust have been accurately removed.

The effect of ICE on silver-based film is devastating. Neither dust nor scratches are removed, but instead fine structured dark areas of a picture turn into black blotches. On the other hand, ICE works without problems with black and white film based on C41 chemistry. As the image of the beluga whale on Ilford XP2 film (C41 process) shows, scratches and dust are reliably detected and filtered out.

Infrared Clean – Scratch Removal in VueScan

In VueScan, scratch detection is handled with Infrared Clean. This requires a scanner with an infrared channel. Unlike ICE, Infrared Clean noticeably affects the texture of the image; however, scratch correction is less effective than ICE.

VueScan is clearly superior to its competitors Nikon Scan and Silver-Fast (which both use ICE) in the following respect: VueScan can scan images as true uncorrected RAW files. In this mode, even after the scan you can still decide whether to apply Infrared Clean, and at what correction level to apply it – Light, Medium; or Heavy. There is no need for a rescan. With Nikon Scan and SilverFast, you must choose the desired ICE correction level before the scan, even with NEF files. In both Nikon Scan and SilverFast, the ICE correction level cannot be changed after the scan.

7.3 Restoring Faded Colors

Nikon Scan: Digital ROC™

Film will age even when properly stored. Over time, the colors in the film will fade noticeably. In the case of slides, since the different color layers don't age equally, they will take on a color cast and become unsuitable for projection. By the way: this is one important reason why film profiles hardly ever seem to match. At different ages the same film looks remarkably different.

Applied Science Fiction has developed the ROC filter for color restoration. Even 20-year old slides can look like new again. ROC removes the color cast and refreshes faded colors. The degree of restoration is selectable; if set too high, colors will look unnatural. Water spots on the film surface (from developing) and other impurities can also become exaggerated.

To check for unwanted side effects beforehand, always generate a preview when using ROC. The ROC filter has a considerable impact on the colors and should be used with caution. The example shown on the following page demonstrates an extreme case. The slide is almost 50 years old and badly discolored. This example shows that even a seemingly hopeless slide can contain some interesting hidden information. ROC corrects typical signs of age very well and improves the image significantly. Current Nikon scanners include ROC. ROC can also be executed after the scan - not with the Nikon scanner software, but with a Digital ROC plug-in for Photoshop at extra cost. The software allows experimenting with different settings without having to perform a time-consuming rescan each time. This is a definite advantage over the solution implemented in the scanner software. For scanning old film material, ROC is a great help.

VueScan: Restore Fading

With Restore Fading, VueScan has integrated a function similar to ROC. However, unlike ROC, the function cannot be controlled with sliders but uses checkboxes instead.

▲ VueScan offers three different levels of scratch and dust removal.

The 50-year-old slide has badly and unevenly faded color layers. Therefore, it has a stark color cast.
(Image: Siegfried Gromotka, ca. 1960)

A correction in Nikon Scan with ROC Level 7 refreshes the faded colors and reduces the
color cast significantly.

7.4 Grain Equalization

Digital GEM

Film is grainy. Graininess can actually enhance the perceived sharpness of an image. Whether this looks good or bad, or to what degree grain looks ideal, is an ongoing discussion among photographers. Film grain was typically optimized to give good results with the wet chemical process of the photo lab; but for scanning, grain can look very unsightly. The visual effect depends on the film and the scanner model. The size of the film grain also depends on the exposure.

Grain in scanned images can look objectionable. Smooth surfaces – especially skin – appear gritty and patchy. The effect is even more pronounced when a high-resolution scan is down-sampled. With the help of GEM, the grain can be made to look smoother. The degree of equalization can be set in different steps. It is a matter of personal taste as to how smooth to make the grain.

The images on the following page were done with Nikon Scan, which lets you set GEM from Level 1 to 4. I deliberately turned off ICE for this example, since ICE also tends to reduce the grain slightly. In practical use, ICE and GEM are used together and their effects are cumulative. It is advisable to perform a test scan to find the best settings. For film with normal ISO speed, reasonable results are achieved with ICE set to normal and GEM set to Level 2.

Since grain size depends on the film material and the exposure, it can happen that different settings for GEM are required for different images on the same roll. GEM is especially useful for high-quality scanners that can record even the finest grain.

VueScan Grain Reduction

The Grain Reduction filter in VueScan is comparable to GEM in Nikon Scan. It has three levels to choose from. VueScan has an advantage over Nikon Scan in that corrections can be performed or changed after the scan, provided the scan is saved in RAW format.

▼ VueScan's filter for grain reduction.

Without grain reduction:
clearly visible grain.

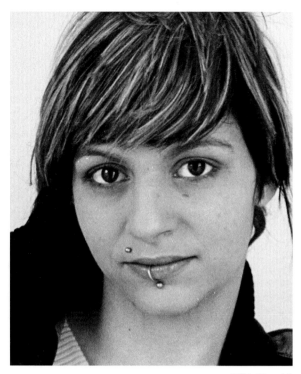

GEM at Level 1:
slightly smoother grain, but still some distracting grain particles.

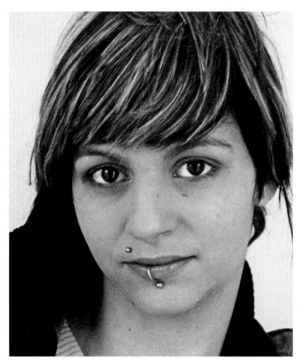

GEM at Level 2:
the graininess is pretty much gone.

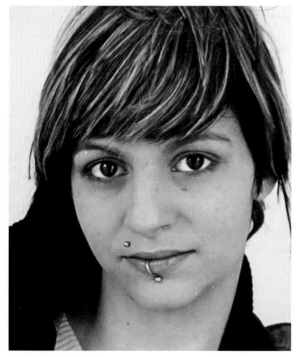

GEM at Level 4:
no more visible grain, but also unnaturally smooth polished skin.

This image was scanned using VueScan with grain reduction. The final image retains the same grain structure as the original.

Even with grain reduction at the highest level the VueScan grain correction is not as agressive as in GEM.

SilverFast GANE

SilverFast GANE has a preview window, which makes it superior to VueScan and Nikon Scan in this discipline. GANE is implemented as an additional filter with three levels. Expert mode allows free choice of strength and threshold. In the RAW mode HDR, the filter can even be applied after the scan, so it should be used only in certain cases and with the strength matching that of the subject.

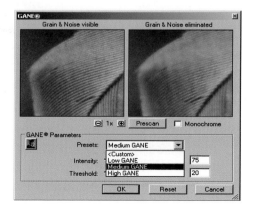

GANE grain removal with preview window. ▶

7.5 Highlight and Shadow Recovery

Digital DEE – Correction for Highlights and Shadows

Starting with Version 4, the Digital Exposure Enhancer (DEE) is integrated into Nikon Scan. It offers automatic exposure corrections for two difficult areas: underexposure in shadow areas and overexposure in very bright areas. Digital DEE can correct the exposure to recover more detail in the highlight and shadow regions. Different regions can be corrected separately. The resulting image can improve drastically.

This feature was introduced with the current generation of scanners (Coolscan V / 5000 / 9000). The same effect can be achieved by manually changing the gradation curves. Since DEE correction can now be performed in Nikon Capture with the D-Lighting feature after the scan, there is no reason to continue using DEE. Unlike DEE, D-Lighting is fully reversible and is not frozen in the NEF files. Starting with version CS, Photoshop has also integrated this functionality with the Shadow/ Highlight adjustment.

Upgrade to ROC, GEM, and DEE Functionality with Photoshop Plug-ins

For ROC and GEM, trial versions of Photoshop plug-ins with the same names can be downloaded from www.asf.com. These also allow scan file corrections after scanning. The plug-in for correcting highlights and shadows is called SHO, but its functions are identical to DEE. All three plug-ins are available in standard and pro versions with extended functions. The free trial versions are fully functional, but the images are saved with a watermark as a reminder to register the software. The advantage of the plug-in solution over the integrated scanner software is that the plug-ins do not require rescanning of the image if the results of the corrections are unsatisfactory.

SilverFast AACO – Autoadaptive Contrast Optimization

SilverFast users can only enjoy the DEE-like function AACO if they have bought SilverFast Ai Studio. The basic version SilverFast Ai does not support AACO at all. AACO offers fewer controls than DEE, D-Lighting, or Photoshop Shadow/Highlight.

Configuring
the Scanning Software

There are a vast number of driver applications for scanners available, each with its own user interface. Nikon Scan is a quasi-standard in this category, so it will be used here to introduce some important tools and options in scanning programs. In SilverFast and VueScan, the buttons look slightly different, but for the most part, the basic functionality is the same. Some settings, such as scan resolution, must be configured before scanning. Others, such as color and contrast adjustments, can be adjusted either immediately in the scanning software or later in Photoshop or other editors.

Contents

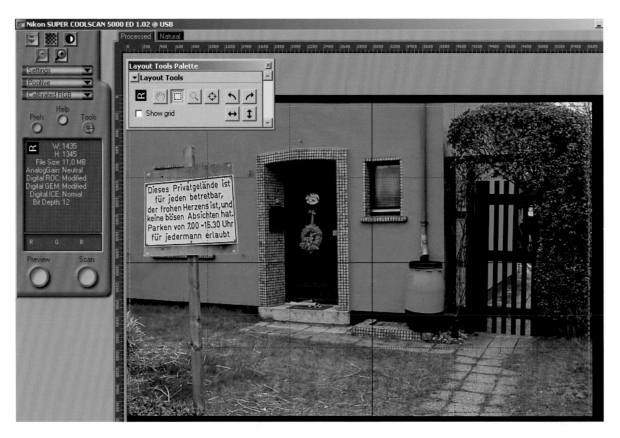

▲ Nikon indicates a mirrored image with
a red "R". When it is black, the image has
not been mirrored.

8.1 Image Orientation

Rotating and Mirroring the Image

An image can be rotated and/or mirrored directly in the scanning software or later during image processing. For scanning, it is important to have the emulsion side of the image facing the correct way. wise, the image will have to be mirrored later. This is not a major concern for rotating an image. In most cases the image editor lets you perform a lossless rotation (90° clockwise and 80° counterclockwise) or mirroring.

Exceptions are JPEG files, which undergo lossy compression with each rotation in some viewers and editors. In some cases, there are problems manipulating the scanned file after scanning: Nikon NEF files, for instance, can no longer be opened in Nikon Scan after they have been rotated and saved in Nikon View or Capture Editor. These problems can be avoided if the film is inserted correctly into the scanner and any rotations are performed directly in the scanner software. At that point, the original is still available for reference for the right orientation.

◀ Automatic frame detection does not always work with dark subjects.

Cropping

Frame gap and frame borders should be trimmed in the preview window before the scan, because the black borders interfere with auto exposure. The question remains as to how to determine the size of the scanning frame. In cropping, you don't want to clip important image information, but you also don't want to set the scanning frame individually for every image. Good scanning software lets you store frame sizes for convenience, especially in batch mode.

The nominal size of the 35 mm format is 24 mm × 36 mm. However, the actual size of the image on the film varies. For cameras with interchangeable lenses, the size may depend on the lens used. For slides, the aperture of the mount is important. Most slide mounts crop the image to slightly less than 24 mm × 36 mm. One exception is the Pro Scan mount from Gepe. The Pro Scan opening is exactly 24 mm × 36 mm.

A slight shift of the film in the mount can further reduce the usable image area. If the ideal scan frame is kept as a standard setting, it will only be necessary to move the position of the frame in the preview window. In Nikon Scan, the crop setting can be saved. This is faster than manually drawing the crop each time. Nikon Scan detects the ideal position of the crop and positions it in the preview. In most dark images, however, the position has to be corrected manually.

Setting Your Own Focus Point

Every good film scanner has an autofocus that can also be set to manual. By default, the autofocus feature focuses on the center of the image, which is fine in most cases. For a flat film, the sharpness will extend across the entire image, provided the image itself is actually sharp.

On the other hand, warped film is not sharp across the entire image, because the scanner's depth of field is usually not sufficient to maintain that sharpness. In this case, it is best not to rely on autofocus, but rather to manually focus on important parts of the image instead, such as on the eyes in the case of portraits.

In my tests, it didn't matter whether there was a contrasty detail or a plain area at the focus point. Contrary to what I expected, the autofocus of Nikon scanners can focus correctly in both cases (unlike the autofocus of an SLR, which only works with contrasty details). I suspect that the film grain provides sufficient detail for the scanner to focus correctly in this mode.

Apart from that, it will always be best practice to ensure that the slide is as flat as possible. This is the only way to get maximum quality out of the source slide.

▶ In shots where the important parts are not in the center, you should reposition the focus point.

Filmstrip Offset

The filmstrip offset must be corrected manually every time a filmstrip is inserted or the automatic detection will not work properly. It is difficult for the automatic feature to separate the frame gap from the image, especially with dark and low-contrast images. This can be corrected by manual adjustment of the filmstrip offset.

8.2 Image Properties

Setting Image Size, Position and Resolution ▸ In the Crop palette you can define exact dimensions for the area to be scanned. The part of the image outside the crop frame will not be captured. If the black borders lie within the crop, auto exposure may give a false result. VueScan and SilverFast perform best in this situation, since they let you set a margin to be ignored. The exact position of the crop is determined by using the mouse. For best image quality, images should be scanned at the optical resolution of the scanner. As a rule, that is always the case for high-quality slides and negatives, with one exception: when negatives from a single-use camera are scanned, it is not necessary to scan at 4000 dpi or even 5400 dpi; 2000 dpi will suffice.

▼ Correcting the filmstrip offset in the prescan window of SilverFast.

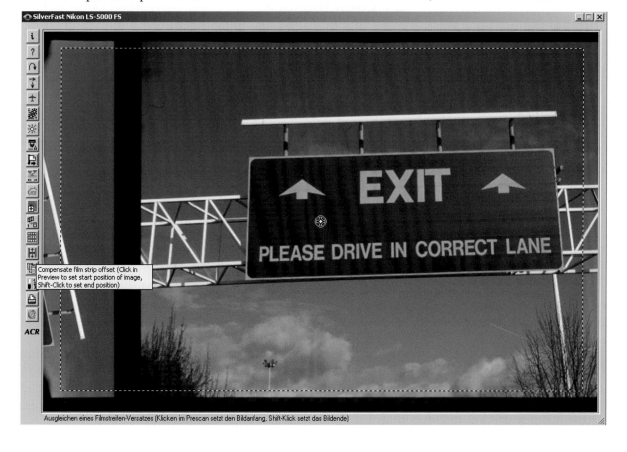

It makes no sense to scan at a resolution higher than the scanner's optical resolution, even if the scanning software offers it. The corresponding interpolation creates unnecessarily large file sizes without any increase in image quality. If you need a higher resolution later (for example, for printing a poster), you can achieve the resolution you need with better control by upsampling in an editor.

Color Depth ▸ Current film scanners support at least 8-bit color depth and in some cases up to 16-bit color depth. For post-processing images, scan at the highest supported color depth. This provides more reserves for processing later. The editor must support 16-bit throughout, though, as is the case with Photoshop CS and CS2. If post-processing is not required, 8-bit color depth will be fine. At any time, a 16-bit file can be converted to an 8-bit file, but not the other way around.

▼ Default setting for color depth is 8-bit only in Nikon Scan. You have to change it manually, if you want to scan in 12-, 14- or 16-bit.

File format ▸ The choice of file format is quite an important decision before scanning. If you are going to produce a high quality digital negative, you should choose either RAW or TIFF. File formats with lossy compression like JPEG are not appropriate for the archiving of slide scans.

8.3 Color Balance

In the Color Balance palette, there are sliders to control brightness and contrast separately as well as the color channels red, green, and blue. These are global settings. To tweak a specific tonal range, use Curves; adjustments to tonal range cannot be done with Color Balance. Brightness and contrast apply to all color channels of the image.

Optimizing the image with Color Balance is quick and easy, but it only allows for relatively rough corrections. For that reason, experienced users prefer the more complicated tuning with Curves. This feature allows much finer color balance and contrast correction.

Brightness ▸ Brightness changes the brightness of the entire image. For example, if many details are overexposed, the overall brightness can be reduced. It is not possible to make a specific tonal range darker or brighter. DEE and comparable functions are available for that purpose.

▼ The Color Balance palette in Nikon Scan.

Contrast ▸ Contrast describes the brightness difference between bright and dark portions of the image. If the contrast is too low, the image looks flat. If the contrast is too high, details get lost and the color distribution in general looks hard and unnatural. Below, the example with the maple leaf is a collage of three partial images. In the middle the contrast is ideal, on the left it is too low, and on the right it is too high.

Color Channels ▸ The color channels allow the red, green, and blue portions of the image to be adjusted. If, for instance, the image has a red cast, the red channel needs to be reduced. The slide below has a minor green cast that has been corrected in the left half.

When the contrast is too low, the image looks flat. When it is too high, it looks unnatural.

On the left, the color cast has been removed by correcting the green channel.

ohne Korrektur   nach Korrektur

The contrast can also be corrected with an S-shaped tonal response under Curves.
The flat original (left half) becomes much more contrasty (right half).

8.4 Unsharp Masking

Digital images, including files from a film scanner, always appear soft in their unprocessed state. Correction filters, such as GEM and ICE, further exaggerate this effect. Unsharp Masking is the best function to sharpen images.

The sharpening occurs by increasing the contrast along edges inside the image. The control parameters for sharpening are Amount, Radius, and Threshold. The goal is to improve perceived sharpness without losing important details and without creating unwanted artifacts, such as halos.

An important factor in sharpening is the resolution of the image. It is rarely useful to sharpen directly during scanning. For example, if the scan goes straight to the printer (i.e., if the digitized image is output in a different resolution from the scan resolution), it is not meaningful to set the "ideal" sharpening parameters during scanning.

In my opinion, Unsharp Masking belongs in post-processing. Even during image processing, Unsharp Masking belongs at the end of the workflow. It is possible with Unsharp Masking to sharpen all color channels together or each color channel individually. Also, simple viewers such as IrfanView provide basic sharpening functions. IrfanView offers only one option for Sharpen, without any adjustable parameters. It is easy to use, but the result is inferior to a sharpening function that can be configured in more detail with Amount, Radius, and Threshold. The USM function in SilverFast is particularly nice: unlike Nikon Scan, the SilverFast USM function uses a preview window for judging the effect of the correction.

▼ USM in SilverFast with its own preview window.

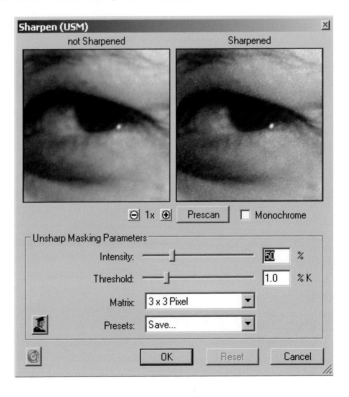

Amount ▸ The slider Amount defines how much the edge contrast will be increased. Ultimately this increase in contrast creates the impression of greater sharpness.

Radius ▸ The slider Radius determines how many adjacent pixels will be involved in edge sharpening process. The bigger the radius, the greater the chance unwanted halos will occur.

Threshold ▸ Threshold defines the minimum level difference between adjacent pixels at which a pixel is considered part of an edge and sharpening kicks in. Setting a high threshold can exclude areas with low contrast. A good example is sharpening eyelashes without affecting the texture of the skin around the eyes.

Sharpening in Two Steps

An alternative to sharpening at the end of image processing is using a two-step approach: first, sharpen mildly after the scan, and then perform final sharpening at the end of image processing. With TIFF files, never sharpen on the scan file - use a backup copy instead. Because sharpening is no trivial matter, countless Photoshop plug-ins, such as Nikon Sharpener Pro, are available to make the job easier.

8.5 Curves

Nikon's palette Curves is somewhat similar to the Color Balance palette in allowing contrast, tonal distribution, and color balance correction. It differs from Color Balance because Color Balance allows only global color changes – all levels and color ranges are treated equally. The Curves function is more powerful since it allows finer corrections, but its use is more demanding. The histogram integrated with the tool helps in evaluating the corrections. Also, Curves allows selective range corrections without unwanted effects on other tones, such as often occur with Color Balance. The color channels can be corrected individually or combined in the RGB channel.

The horizontal axis of the diagram represents the input values; the vertical axis, the output values. The "0" on the left is the blackest part of the image, and "255" on the right is the maximum possible brightness. In the histogram, the vertical axis displays the frequency with which a given level occurs in the image. If a level occurs frequently, then the bar for that level is high; if the level is rare, the bar stays low. The vertical scale for the histogram is not fixed, but is adjusted flexibly to have the tallest bar always fit within the graph of the curves. The numerical values on the vertical axis define the black point and white point on the output side.

Initially, the curve is a straight line, which means that input and output values are identical. Pushing the curve increases the brightness level of specific tonal ranges. Moving the mouse over the image displays a point running along the curve to indicate the level value of the pixel currently under the cursor.

Difference from Curves in Photoshop

Nikon integrates the histogram with the Curves tool. In Photoshop, Levels and Curves are two separate tools, with no histogram display in the background of the Curves graph. A similar function is available with the live histogram of the Histogram palette; with the Curves window placed next to the histogram, the histogram levels change as the curve is pulled.

Working with Histograms

The histogram helps assess the effect curve correction has on the distribution of tones in the overall image. The histogram can show the distribution of levels not only of the full image, but also of a selected crop. The histogram displays the frequency (y-axis) of pixels for each level (x-axis) within the image. This type of display provides a very helpful tool for image processing. For example, a distribution spread evenly across the whole scale from 0 to 255 indicates a correct exposure with balanced levels. These 256 steps are based on 8-bit RGB standard. Even if your are processing 16-bit images the scale still has 256 steps. But this is just a simplified display; internally, of course, the program uses 65,536 different steps as needed for 16-bit image processing.

◀ ▼ The histogram of an underexposed image shows – as expected – a disproportionate concentration of dark levels on the left.

▲ For normal images, a bell-shaped histogram indicates a balanced exposure. Shifting the black and white point increases the contrast.

When adjustments are made in Curves, the histogram does not update in real time, even though the changes affect the histogram. The Histogram button in Nikon Scan is used to display changes. It also functions as a toggle to show "before-and-after" values in the histogram. In Nikon Capture, this button offers a live visual update of the histogram. Unlike Photoshop, Nikon Capture does not offer superimposed histograms for a before-and-after comparison.

There are no fixed rules for how a good histogram should look; the subjects and their level distributions differ too greatly. Gaps in the histogram after corrections are always a concern. They point to a loss of detail due to shifted levels.

Before/After Histogram Button and Auto Contrast

Using the Before/After Histogram button displays the effects of the correction. With one mouse click, the function provides a comparison of the updated graph with the original one. Unfortunately, in Capture Editor it is hard to tell whether the "before" or "after" histogram is being displayed. The Nikon Scan interface is clearer, and it is easier to determine the proper settings. Using the Auto Contrast button lets the program

◀ On the left is the original, and on the right is the same image after stretching the levels. The result is more brilliant colors.

◀ ◀ Uncorrected histogram: the entire level range is not used and the image is flat.

◀ After manually correcting the black point and white point, the levels are optimized.

make an automatic contrast adjustment. In my tests, the results of performing contrast corrections were in many cases unsatisfactory. Manual corrections seem to be better. There is no way to undo Auto Contrast except by completely resetting all curves. This can be irritating if the curves have been changed extensively; those changes will all be gone as well.

Black Point and White Point

The Black Point is the darkest pixel in the image, indicated by the value at the left end of the histogram. The White Point is the brightest pixel in the image, indicated by the value at the right end of the histogram. Usually, the Black Point is at 0 and the White Point is at 255. However, in many images the darkest point is not deep black and the brightest point is not pure white. This means that a part of the dynamic range is not being used. Manually moving the two sliders to the respective ends of the histogram will use the entire dynamic range.

The fastest way to adjust levels is to use the RGB channel, although it is not the best way. It is more effective to make the effort to adjust the levels for each color channel separately. This not only makes best use of the tonal palette, but also compensates for most color casts (with the exception of high/low-key shots).

Another way to adjust levels is to use the eyedropper tools Set White Point and Set Black Point. First click on the eyedropper to activate it, and then click on the desired part of the image. For example, to set the white point, click on the pixel that is considered the brightest. Areas with even brighter pixels will be washed out. The sliders for black point and white point can be set separately for each color channel. If the eyedroppers in the RGB channel are used, then each color channel will be adjusted automatically. This is also a good way to fix color casts, provided suitable black or white points can be found in the image (which is not easy for every subject).

Setting the Gray Point

▼ The gray point slider before (top image) and after (lower image) the correction.

The gray point lies between the black point and the white point. By default, it is right in the middle. Sliding the gray point to the left makes the image brighter. Sliding to the right makes the image darker. Unlike the global brightness adjustment in Color Balance, this correction does not affect the endpoints of the curve. Black point and white point remain unchanged; only the midrange is affected.

With the gray value, the gamma changes. One can slide the gray triangle or type a number in the field below it. Alternatively, the gray eyedropper may be used to assign a certain image area the value "neutral gray"; each color channel is then corrected separately. This is a very good method for removing any color cast, but only if there is a suitable neutral gray area in the image.

There are not, however, many images that genuinely have such neutral gray values. For this reason, manufacturers such as Novoflex and Kodak produce standardized gray cards that reflect exactly 18% of the light. Such gray cards allow digital photographers to adjust white balance manually while shooting, thus virtually guaranteeing accurate color rendition. The situation is more complicated when scanning, because the film material itself can produce color casts, and because the general degeneration of film material over time produces further unavoidable color shifts. Digital photographers have an easier time of things, as the image sensor in a digital camera remains the same over time, and digital images don't age.

Ideally, you should include a gray card in every scan you make, to enable you to adjust color balance accurately at a later stage. This is, of course, extremely impractical, and so the general use of gray points in scanning situations remains limited.

uncorrected

corrected

8.6 LCH Editor

The LCH Editor performs image corrections according to the concept of the LCH color model. Here you can adjust colors separately through their components: Lightness (Luminance), color saturation (Chroma), and color tone (Hue). The LCH Editor gives the photographer a powerful tool for optimizing images. In Nikon Scan, the LCH Editor is functional only if Nikon color management is enabled. Otherwise, the LCH Editor options are grayed out and cannot be used.

Lightness ▸ This option allows adjustments to the brightness of the entire image. Colors are affected only indirectly; a very bright image automatically looks paler.

Chroma ▸ The Chroma Editor allows changes to the color saturation of the image. In the example on the following page, using the Chroma Editor makes the colors in the parrot's plumage more vibrant. Unfortunately, scanning degraded the colorful image somewhat, but with the LCH Editor the quality of the color could be recovered to a degree.

The color saturation is controlled by the gray point slider in the middle. Moving the slider to the right makes the image become less saturated; sliding it to the left makes the colors more saturated.

▲ The original image (on top) is a bit too dark. Correcting the gamma with the gray point slider makes the mid-tones brighter (lower half).

▾ Adjusting brightness with the LCH Editor in Nikon Scan.

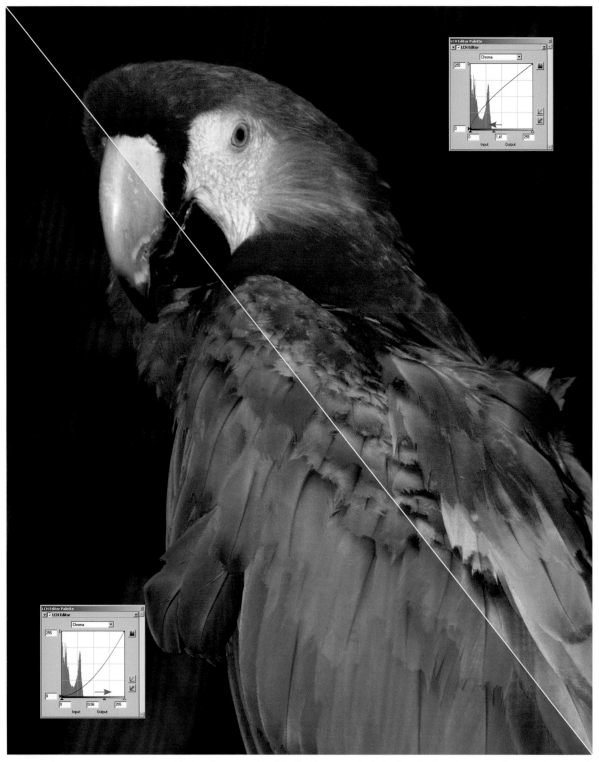

Moving the gray point slider for Chroma to the left will make the colors more vibrant and saturated.
Sliding it to the right makes the image less colorful.

Chroma in Capture

In Nikon Capture, the Chroma feature looks different; it is not shown as a diagram with a curve. Instead, it is shown as a spectral diagram with a horizontal line. Color saturation is changed by grabbing the arrow and pulling the line up or down. The effect is the same as in Nikon Scan.

In Capture, color saturation is increased by pulling the line upwards. ▸

Hue ▸ The Hue Editor also works with a spectral diagram instead of the more common curves diagrams. The background shows all the colors from the color wheel along the horizontal axis, which means the full 360°. The vertical axis toggles between 60°, 120°, or 180° of the color wheel. With 60°, there is less of a shift but a higher resolution.

At the outset, the input hues are identical with the output hues. The numerical values for input and output are shown at the bottom. Grab the straight line at any point and drag it upward or downward to assign different values to the input value for the selected range of hues. The width of the change is adjusted with a slider on the bottom left to allow some fine control. This way, it is possible to choose whether only one particular color or also its neighboring colors are affected by the change.

Example:
Change the Color of the Parrot's Head with the Hue Editor

Activate the Hue Editor by clicking on the diagram. First, the hue of the original color must be found – here, the red color of the head. Moving the mouse cursor over the head shows the hue as a little point on the horizontal axis and also as a numerical value in the status line at the bottom. Clicking on this location on the axis sets a marker. Using the mouse or the arrow key pulls this marker into a different color region, until the color of the head turns the desired turquoise. Usually this change will also affect neighboring color regions, but this effect can be fine-tuned with the Width Slider. The width must be set narrow enough so that only the desired color range changes.

On the hue curve, the input value for red is pulled upwards into the turquoise region. The formerly red head of the parrot has changed color dramatically.

8.7 Analog Gain

Analog Gain of Nikon Scan (Lamp Brightness in SilverFast) is usually taken care of automatically by the scanning program as part of exposure control. It is a matter of exposure time and not the brightness of the light source, as the name would seem to indicate. For film with normal exposure, this function is not really necessary and the exposure may be left at its default setting. For underexposed or overexposed pictures, manipulating Analog Gain produces better image results.

Analog Gain directly controls the scanner hardware. It governs exposure time, i.e., how long the scanner lamp shines through the film. The same effect cannot be produced later in image processing; more light must be physically projected through the film.

Nikon Scan simulates the effect in the preview window to give a point of reference. However, only performing the proper scan would show the exact result of adjusting exposure. Increasing Analog Gain makes sense only if the monitor is calibrated. Often, a scanned image looks too dark simply because the monitor is not adjusted properly. If one tries to compensate for poor monitor adjustment with Analog Gain, the scanned files will not be usable on a calibrated monitor.

The effects of Analog Gain are fixed in the image file. Since Analog Gain represents change at the hardware level, a poor setting cannot be fixed later in Photoshop. A negative side effect of using Analog Gain is an increase in image noise. The simplest way to work with Analog Gain is with the Master slider (all three color channels together). In most cases this works fine. It is also possible to tune each color channel separately. However, finding the ideal settings for tuning can be tedious.

▼ When Brightness or Curves can no longer extract more detail from an underexposed image, Analog Gain can help.

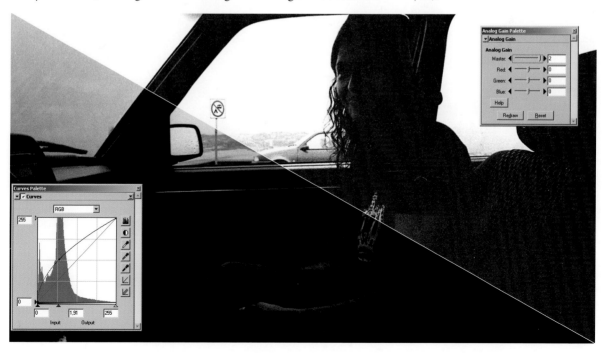

For overexposed images, reduce Analog Gain to lessen the exposure time. This may salvage image portions that would be blown out by the scanner light on the standard setting. Analog Gain does not work miracles; it merely helps to recover some detail in poorly exposed images. The results of using Analog Gain will never match a properly exposed shot. Best practice is always to ensure correct exposure at the moment you take the picture. If that is not possible, Analog Gain is at least a handy makeshift to rescue some details.

Analog Gain vs. Multi Exposure

▼ In the overexposed slide, many details are blown out. With reduced Analog Gain, the sky shows more detail. The drawback is a strong color cast.

Multi Exposure produces an effect that is similar to Analog Gain, but in contrast it expands the dynamic range of the scanner on the whole, (which is the better technology). My advice would be: if your scan program supports Multi Exposure try it out first. If it does not have the desired effect, then try Analog Gain as a last alternative.

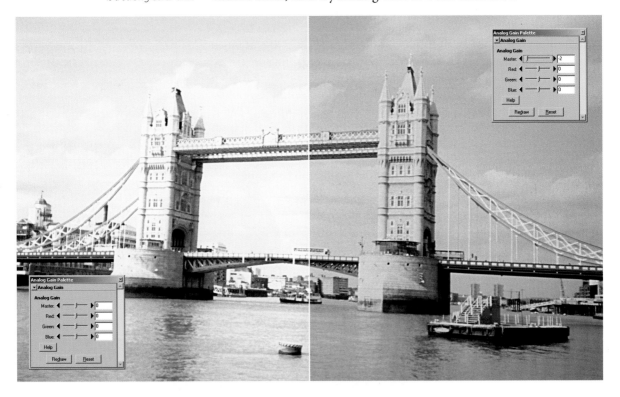

Nikon Scan

Nikon scanners include Nikon Scan software. This software will be described here as a member of the TWAIN family of drivers, which are typically included in a scanner purchase. The tight integration of Nikon Scan with the RAW editor Nikon Capture NX2 makes it especially interesting to photographers. Currently, no other program offers ease of use comparable to Nikon Scan at this level.

Contents

9.1 Installation

Nikon Scan and program updates can be downloaded from the Nikon support website free of charge. The software supports Nikon scanners only. It should be noted that the current versions no longer support all old Nikon scanners. Nikon Scan 4 is compatible with:

▸ Coolscan IV + V ED
▸ Super Coolscan 4000 + 5000 ED
▸ Super Coolscan 8000 + 9000 ED

Nikon Scan can be used either as a fully functioning stand-alone application or as a plug-in for image editors such as Photoshop. The plug-in solution is convenient for situations in which the scanned file is processed immediately with an external image editor. However, during scanning in the plug-in mode, the editor is locked. It is better to save the image files directly to disk and edit afterwards. When Nikon Scan is run in stand-alone mode, images can be edited in parallel with Photoshop or Capture – provided there is sufficient computing power to do so.

The plug-in mode has a limitation you should know. Nikon Scan can transfer color space information to the image file only in stand-alone mode. The TWAIN interface used for Plug-in mode cannot communicate the color space of Nikon Scan to the application.

▼ Almost all fine adjustments can be configured in the Palette Tool.

9.2 Look and Feel

The layout of Nikon Scan is visually simple but highly functional. It takes some time getting used to, but after familiarization, it is easy to use. When the application starts, the icon Open Twain Source opens the actual scan window. This only works if a scanner is already connected and recognized by the computer system. If not, an error message indicates that no active device has been found and the scan window cannot be opened. In this case, there is no need to restart the program; just switch on the scanner, wait for a short while, and then open the window again.

Layout of the Scan Window

Control Area

In the control area, there are three drop-down menus: The menu Settings lets you save and recall configurations for the current scan. The Film Type menu lets you select among settings for color negative, black and white negative, Kodachrome slide, and non kodachrome slide film. No further film profiles can be attached; you have to take care of color correction for the different films yourself. In the Color Model menu you can choose between Grayscale and Calibrated RGB. You should use Calibrated RGB when scanning in NEF format. NEF files should not be scanned in grayscale mode, because neither Nikon Capture NX2 nor Nikon View support it, which makes post-processing difficult.

▼ In grayscale mode, you will face compatibility problems later.

Information Panel

The information panel in the scan window clearly shows the most important settings, readouts, and status for the currently selected image. A yellow warning sign indicates when image corrections configured in the scan program do not show up in the preview window yet. This can be ICE, ROC, GEM, Analog Gain, Scan Image Enhancer and DEE which are not yet taken into account in the preview. Only the repeat of a preview scan allows a realistic evaluation of the image corrections; then the yellow warning signs will disappear. The information panel is quite handy for a quick glance at the most important settings, but for detailed information the Palette tool is the better choice.

▼ The most important settings at a glance.

▲ For a comfortable comparison, you
can always go back to the uncorrected
source scan with one click.

Preview Area

In the preview area, the image can be viewed with the two tabs Natural
and Processed. You can toggle between them either by clicking with the
mouse or by using the keyboard shortcut `Ctrl+F5`. Whereas the Natural
option shows a preview without any image corrections, the Processed
option shows a preview that includes image corrections, which allows a
direct before/after comparison. Whether the preview matches the actual
scan, and to what extent, can be configured under Preferences with Preview
Settings. If Digital ICE is activated, the projected effects of scratch
removal can be checked in the preview as well.

Title Bar ▸ The title bar displays information about the connected scanner
and which version of firmware it is running.

Thumbnail Drawer Tab ▸ The thumbnail drawer tab is somewhat hidden
to the right of the control area in the scan window. Open the drawer
either by clicking on the tab or using a keyboard shortcut: the function
key `F4` opens it and `F9` closes it again. (Note: shortcuts work faster than
the mouse.) The thumbnail drawer automatically closes after double-
clicking on a thumbnail to start the preview.

The small size of the thumbnails and their distorted representation
do not allow a realistic evaluation of the image. However, the represen-

◀ Thumbnails à la Nikon Scan: the faces are slightly garbled and the person looks different in each picture..

◀ The best spot for the tool chest is next to the scan window, not in front of it.

tation is good enough to sort out obvious exposure errors and to correct the filmstrip offset. Nikon Scan does not allow thumbnails to be saved to disk, although it is possible to do so with VueScan and SilverFast.

Tool Palette 1

The Tools button lets you open Tool Palette 1 in a floating window that provides all tools for image processing. With this palette, you can build your own tool chest. In general, you do not need all the functions available in Tool Palette 1. You can often work faster with a palette window tailored to your own needs.

You can move a tool title bar (such as Crop) by clicking on it and dragging it by holding down the mouse button. This way you can rearrange the sequence of Tool Palette 1 or even create a new palette by dragging the tool outside the existing palette. Palettes with a single tool receive a palette title with the tool name – in this case, Crop. For multiple tools in the palette, tools are numbered sequentially.

Unfortunately, it is not possible to name the palettes; Nikon Scan gives them names automatically. Also, Nikon Scan does not let you set the palettes to always stay in the foreground. This can get messy when the scan window and palettes overlap, which is often the case with a regular size monitor. When you click on the image, the palettes disappear behind

▼ Trimmed down Tool Palette 1: not all default functions are needed.

the scan window and need to be restored by clicking the Tools button. Capture has shortcuts such as ⌥Alt+1 (where the number stands for the palette number; in this case, Tool Palette 1), but these shortcuts do not work for Nikon Scan. If you have a large monitor, it is best to place the palettes next to the scan window.

Scan Progress

The Tools button is also used to bring up the Scan progress window. This is a floating window with tabs for Queue and Log. The Queue tab lists remaining scan jobs, which can be useful for batch scanning. The Log tab lists completed jobs, which helps keep track of a sequence of steps. Unlike VueScan and SilverFast, Nikon Scan does not have an acoustic notification for the completion of a scan.

Preview Area

The larger the preview, the more effectively you can work with it. Therefore, it is best to first enlarge the window, so that Nikon Scan will generate a correspondingly large preview. If the window is enlarged afterwards, the preview becomes pixilated, which can only be fixed with a new preview. Pixilation can sometimes occur when an image is rotated (for example, from landscape to portrait). If you are satisfied with small thumbnails, you would expect faster previews – higher resolution takes longer to scan. This is not quite the case; in actual tests with a Coolscan 5000, it did not matter whether the preview was filling the screen or was as small as a slide mount - the scanning time was the same in both cases. Thus, it is best to keep the preview window at maximum size; it can be reduced any time. Nikon Scan does not allow thumbnails to be saved, only the actual scan. The same is true for SilverFast - only VueScan has no problem saving thumbnails. Saving thumbnails is a nice feature for playing around with the program, but actually you will not miss it during scanning very much.

9.3 Saving Configurations

Coming up with optimized combinations of various configurations for different purposes can be extremely time-consuming. Therefore, it is very useful to be able to save settings at different levels so that you do not have to repeat those settings before each scan. If you know the different ways of saving settings and apply them appropriately, you can save time and avoid tedious, repetitive tasks in the future. Especially for scanning in batch mode, this knowledge is quite helpful.

The Settings Menu

The Settings menu allows you to save and delete configurations. Some of the settings are:
▸ Scanner model
▸ Media type: color negative, black and white negative, slide
▸ Crop
▸ All options in the tool chest

Unfortunately, Settings does not allow the selection of individual sub settings – all settings are saved collectively. If you get lost in the various configurations, you can always go back to square one with Reset to Factory Defaults. But in this case all configurations you have made in the meantime are set back, too.

If the preview window doesn't display ICE settings you have made, this is due to the Prescan settings. These determine whether the scanner samples ICE during prescan. If you deactivate ICE in prescan, you save valuable time during every prescan sequence.

Store and Delete Settings

The Save Settings option under Settings lets you add newly created settings to the bottom of the menu as a menu option with a name of your choosing. If you save many different settings, the menu can become unwieldy. You can remove obsolete settings from the list with Delete Settings.

Set User Settings

Set User Settings lets you define the current settings as user settings. This allows you to make a set of customized settings for your personal needs. All settings can be reset to these values with Reset User Settings in the Settings menu. Additionally, there is a Reset to User Settings in most tools of Tool Palette 1. This is very useful should you get lost in the jungle of different options.

Import and Export Settings

In order to keep the drop-down Settings menu from becoming too long, configurations can be exported to a .set file and imported again when needed. Only current configurations can be exported. To export a configuration (using the option listed in the drop-down menu under Settings), you first have to select the configuration and make it current.

▶ Customization: defining user settings.

Saving Configurations in Tool Palette 1

The Settings menu only lets you save all settings together. This includes settings for the tool palette as well. Individual settings can only be set directly in tool palettes such as Tool Palette 1. Functions in these palettes follow the same logic as functions under the Settings menu. Configurations can be stored, set for display on the drop-down menu of the tool, or imported and exported. However, not all tools in the palettes are supported. At the tool level, the triangle at the right end of the title bar gives access to the Settings. Tools that are not marked with a triangle do not support saving, importing, or exporting of tool parameters. Copy Settings to Clipboard is another handy function available with a few tools. Once on the clipboard, the settings can be applied to an image through the menu option Paste under Edit.

▲ Export of the ICE configuration

Crop ‣ The settings for image size, resolution, and position of the cropping frame can be exported to an .ncc file.

Layout Tools ‣ Layout tools do not support the export of any settings at the tool level. Any configurations generated in layout can be saved only through the drop-down menu under Settings in the scan window.

Curves ‣ The settings for curves can be saved to an .ncv curve file and retrieved later.

Information ‣ This tool displays various image information – saving and exporting settings are not possible.

Color Balance ‣ Color balance files can be saved in .nca format.

Unsharp Mask ‣ Settings for the tool Unsharp Mask have their own file format, .num.

LCH Editor ‣ The LCH curve data can be saved to .nlv files.

Digital ICE Advanced ‣ Settings for Digital ICE Advanced have their own file format, .nla.

Analog Gain and Scanner Extras ‣ Neither Analog Gain nor Scanner Extras allows saving or exporting settings at the tool level.

9.4 Preferences

The Preferences option may be opened either with the Prefs button or with the shortcut `Ctrl`+`K`. Preferences contains configurations to be applied to all scans globally.

File Locations ▸ File Locations shows the path to temporary files. These files should be located on a separate partition with plenty of free space and without too much fragmentation. If the partition is badly fragmented, scanning slows down. If possible, the temporary files and the scan files should reside on a different partition from the operating system, preferably on a different hard disk.

Single Scan ▸ The menu item Single Scan contains the settings Autofocus, Auto Exposure for Positive Film, and Auto Exposure for Negative Film. For manual focusing, Autofocus should be deactivated. Alternatively, you can set the focus point with the focus tool by leaving Autofocus active. The most important setting in the menu Single Scan is Save to Disk. The

▸ The basic setup is done in Preferences.

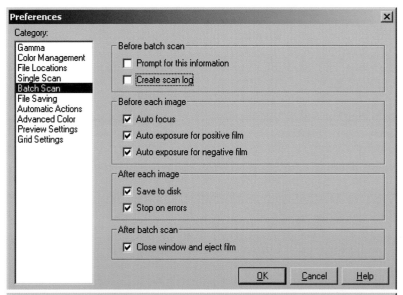

▸ Save to disk is very helpful for batch scanning.

image file will be saved directly to disk only if this box is checked. If not, a window with the image will open and you will have to save it manually.

Batch Scan ▸ The settings under Batch Scan, for the most part, match the ones under Single Scan. There are three additional options. The option Prompt for This Information shows a selection menu before the batch scan. When prompted, you can double-check the set options before launching the batch scan. This helps getting used to the program in general. Once you have developed a routine, the function becomes superfluous and you can turn it off to speed up your work.

The option Create Scan Log generates a log. This can be useful if problems occur during the batch scan and you want to find out how much time is being spent on a single scan. (Note: this option can be deactivated for daily use; the scanner runs fine without the log.)

The option Stop on Errors stops the batch scan in case of a malfunction and allows you to fix the scan right away.

File Saving ▸ The menu item File Saving lets you define a default file format for saving. This is simply a default that can be changed during the scan. The option Last Used is even more practical. If you work with different formats, once you choose a specific format, that one remains the default until you change it again.

Automatic Actions ▸ With the function Automatic Actions, you can automatically generate a preview and thumbnails when a filmstrip is inserted. This saves a few mouse clicks. Thumbnails are needed to identify and correct filmstrip offset. Activating one of the options under Thumbnail when film is inserted will start the function automatically. However, simply inserting the filmstrip will not activate the scanner; the scan window must be already open.

Whether the option Perform Autofocus When Focus Point Is Used is helpful is a matter of taste. To engage it, you first have to set the focus point in Tool Palette 1. It is actually faster to initiate a manual focus with the Autofocus icon in the option area of the scan window. To do that, hold down the $\boxed{\text{Ctrl}}$ button, click on the Autofocus icon, and then click on the desired location in the image. The system will focus immediately without going through Tool Palette 1 first.

It is worthwhile checking the Multiple Image Adaptor if all of your images are of excellent quality. Often the thumbnails can be presorted to avoid the time-consuming task of generating previews for poor images.

Ejection of the Film
Nikon Scan offers the option to automatically eject the film-strip after batch scanning to indicate the end of the job. This function can be set with Views ▸ Preferences ▸ Batch Scan ▸ Close Window and Eject Film. When the job ends, the scan window closes as well and must be reopened for the next scan.

▲ All formats supported by Nikon Scan.

Advanced Color ▸ The functions under Advanced Color for RGB and Grayscale let you define the White Point, Gray Point, Black Point, and the default for the Auto Contrast Calculations. These functions should be considered for fine-tuning only; the standard settings are quite usable. Apply changes at this place only if your specific system does not deliver good results with standard settings.

Preview Settings ▸ Functions under Preview Settings allow for AutoFocus and Auto Exposure for Positive/Negative Film. If you prefer the highest quality even for the preview, you can activate Multi Sampling and Digital ICE as well. Both prolong scanning time considerably and are not really needed unless you want to tweak the preview for maximum quality. All these settings are irrelevant to the result of the final scan; they only affect the preview. Nikon Scan supports Multi Sampling only for the models 4000 ED, 5000 ED, 8000 ED, and 9000 ED. Not only is it not needed for the preview, but you will not be able to see any quality gain from Multi Sampling in the relatively small preview frame. For Digital ICE it is different. If you use Digital ICE and still see some foreign matter such as dust on the image, you should clean the original one more time before scanning.

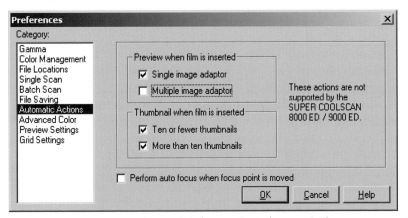

▲ Automatic index scanning makes sense in almost every case. For preview scanning this is generally not the case.

▲ The fewer boxes checked, the faster the preview will be generated.

Grid Settings ▸ You can put grid lines over the preview with Tool Palette 1▸ Layout Tools ▸ Show Grid. To configure the grid, select Prefs ▸ Grind Settings. You can use your imagination to freely set grid spacing and line color. It can be handy to define the grid spacing in pixels. In the sample image below, grid spacing is set to 1535, which corresponds to a 6" wide print at 300 dpi. An enlarged crop for a 4" × 6" print in itself covers over a quarter of the image area. The grid helps image flaws become more noticeable, such as a tilted horizon; but a correction is possible only later in Capture NX2 or Photoshop. Nikon Scan has no tools for straightening.

To make these settings the default, select the option Show grid for new windows. With this option, every new review will show a grid. If the grid is bothersome, it can always be turned off in the current window with Layout Tools.

▲ A grid with a particular spacing can help judge the potential for enlarging crops.

9.5 Color Management

The Nikon color management system is supposed to guarantee a consistent color reproduction from scanner through screen display all the way to printed output. If you want to take control yourself, you can deactivate the color management function. When color management is deactivated, the color space is not applied to the image. You get unprocessed RAW data from the scanner. If color management is activated, then a color space transformation will be performed, even when NEF files are used.

Before I go into too much detail, I can assure you that Nikon has integrated a very powerful color management system into the software. Unfortunately, though, they seem to have missed the most important point: namely, that the image color data have to be read accurately by the scanner in the first place.

Input Profiles / Scanner Profiles

The input profile used by Nikon Scan depends on the scanner model and the film type. It is automatically selected by Nikon Scan and cannot be overridden. Operator error is thereby eliminated. Apart from selecting a scanner and a film type, there are no other ways to configure input profiles. It is not possible to generate your own ICC profiles and apply them to your own scanner. Nonetheless, Nikon Scan comes with working scanner profiles for all supported scanners. SilverFast and VueScan do not include ready-to-use scanner profiles, but unlike Nikon Scan, they let you use your own ICC profiles.

With the Nikon solution, you cannot use ICC profiles to correct changes in the color characteristics of the scanner, which could be caused by an aging light source or by manufacturing tolerances. This is a shortcoming of the Nikon scanner software.

Monitor Profiles

In Nikon Scan, the suitable ICC monitor profile is set with Prefs ▸ Color Management ▸ Monitor, but only if color management is activated; otherwise, the tab is grayed out. The ICC profiles chosen here affect the display of the image in preview and during image editing in Nikon Scan. Using the proper monitor profile will ensure that you are correcting the color cast of the image, not the color cast of the monitor. The profile affects only the color representation in the image window and scan window of Nikon Scan; other programs remain unaffected. Nikon Scan already includes a number of standard profiles, but using an ICC profile specifically made for the monitor is always preferable.

Selecting the option Use factory default monitor profile adopts the Windows default configuration, thereby making a custom configuration in the scanning software unnecessary - provided that Windows already utilizes a suitable profile.

Color Space Profiles

Nikon Scan supports a number of color spaces that can be set under Preferences ▸ Color Management under the tab RGB. For most applications, Windows users will choose either sRGB or the wider color space Adobe RGB.

Nikon scanners can read more color information than even that represented by the color space Wide Gamut RGB (compensated). That's why Nikon offers another item on the list with the color space Scanner RGB. This option eliminates conversion to the selected color space, along with its associated color losses. It deals with the entire color range of the scanner's RAW data.

Unlike the case in which color management is completely disabled, using the profile Scanner RGB offers the possibility to work with the LCH Editor and use the Unsharp Mask function. These functions would otherwise be unavailable when color management is deactivated in Nikon Scan.

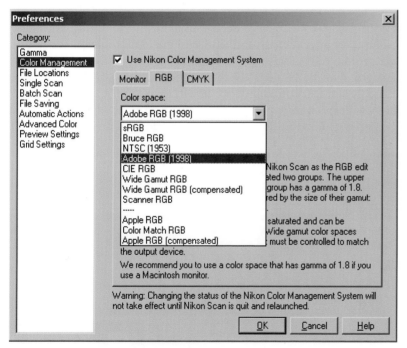

▲ Color Space is configured in Settings.

Setting a Monitor Profile as Windows XP Standard
By right-clicking on the desktop and selecting from the menu Properties ▸ Settings ▸ Advanced ▸ Color Management, Windows lets you define a monitor profile as the default.

*Special Case Photoshop: Nikon
Color Management Is Ignored*

When scanning from inside Photoshop, the Photoshop settings determine the color space assigned to the image file, not the settings in Nikon Scan. The same applies to other image editors that can use Nikon Scan in plug-in mode. It is the setting of the image editor that is important.

There are two preconditions that must be met before a selected color space can be embedded in the image file. First, the chosen file format has to support color profiles, which is the case for TIFF and NEF. Second, Nikon Scan has to run in stand-alone mode. In plug-in mode the TWAIN interface does not allow the transfer of color spaces.

Besides RGB, the color model CMYK is also available. Internally, the Nikon scanners work in RGB mode, but they can convert to CMYK profiles. This function is of little interest for home users, but of great interest for prepress.

The topic of color management in Nikon Scan is discussed in great detail in the guide to Nikon Scan 4 color management. This document is available in PDF format for download from the Nikon website. The guide can also be found in the user manual.

Film Profiles

Nikon Scan includes the film profiles Positive, Negative (color), Negative (mono), and Kodachrome. It is not possible to choose profiles for more specific film types or to generate and apply your own profiles. Since every film has its own color characteristics, there is no setting which suits all films equally. VueScan and SilverFast use film profiles to convert negatives to positive image files as true-to-color as possible. These profiles contain color corrections for the individual film types. On the other hand, Nikon relies on a hardware-based conversion; it is supposed to do away with special film profiles. Still, individual color adjustments can be carried out and saved in Tool Palette 1. To do that, the ideal correction

◀ ▼ VueScan offers many film profiles;
Nikon Scan, only four.

for a film is determined from a test image and then applied either directly during the scan or afterwards to the file. This method is more efficient than correcting the colors for each single image.

In practice, the Nikon MAID libraries often deliver better analysis of the original than the preset film profiles included with VueScan and SilverFast. You won't miss the additional film profiles in Nikon Scan, though.

9.6 Documentation

The Nikon Scan documentation is very good. The online help, which can be started with Help, is rather basic, but there is an extensive reference manual available in PDF format.

The latest versions of the manual can be found on the Nikon support website. Every function is described in the approximately 140-page manual in great detail. The manual also takes advantage of the features of the PDF format: Hyperlinks allow convenient navigation between scan functions. Even without using Nikon Scan, the manual is quite interesting to read to get a general overview of scanning techniques.

▼ The Nikon reference manual is clear, vivid, and informative.

9.7 Conclusion

The proprietary NEF file format and its integration into Capture and the viewer is an exclusive feature of Nikon Scan. The disadvantage of the otherwise excellent Nikon Scan support for RAW data is that the ROC, DEE, GEM, and ICE filters are firmly applied to the image. Regarding this aspect, VueScan is offering a better solution. But when the entire workflow is considered, Nikon Scan offers the most convenient way to work with RAW files. No other solution lets you generate RAW data and then view, edit, and save it in different versions.

Multiple copies are not even necessary; a single file can do it all. On the other hand, Nikon implemented color management only halfheartedly. A big disadvantage of Nikon color management is that you cannot apply your own ICC profiles to the scanner. The software includes profiles for each supported scanner, which themselves cannot be modified. The competition is doing better in this regard: SilverFast and VueScan both support color management throughout.

All in all, particularly because it is free of charge, Nikon Scan is a serious alternative to SilverFast and VueScan. This is, however, only true for negatives. Nikon Scan is less suitable for use with slides—the missing IT-8 calibration function means scans end up with visible color casts in most cases. Nikon has basically halted development of its scan software, so this situation is not likely to change in the foreseeable future.

VueScan

With VueScan, Ed Hamrick has written a lean but nonetheless powerful scanning tool for Windows, Mac OS, and Linux operating systems. Neither SilverFast nor Nikon Scan support Linux, so VueScan is the only option for this operating system. Unlike its two competitors, VueScan is purely a scanning program with only some rudimentary editing functions. This is not a disadvantage if you use a good image editor for postprocessing. The program offers powerful features with full and impressive RAW support. Also, unlike Nikon Scan and SilverFast, extra, expensive programs are not needed for editing RAW data.

10

Contents

10.1 Purchase and Installation

VueScan is available in Standard and Professional editions, which cost $39.95 and $79.95 respectively. The software is available only through download from Ed Hamrick's website, www.hamrick.com. It is advisable to back up the installation files of the version for the purchased serial number, because the website has only the newest versions available for download (which may require a different serial number). The purchase price of the Professional Edition includes updates; otherwise, for updates, you have to request a new serial number. VueScan only comes with an English user interface.

Professional Edition Recommended

Unlike the professional Edition, the standard Edition can neither generate RAW data nor does it support ICC profiles and IT8 calibration. Therefore, for archiving film material, the Professional Edition is the better choice. VueScan currently supports 750 different scanners. The license covers any number of scanners you own. (SilverFast is less generous: you need a separate license for each scanner, because there is a program version for each scanner type.)

With a size of less than 3 MB, VueScan installs in a snap. The installer creates the directory \VueScan on the local hard drive and a shortcut on the desktop for convenient launching, but does not generate any entries in the start menu. To uninstall, simply delete the folder and the shortcut. VueScan is a stand-alone program, but it can automatically open the scan file in an image editor for further processing.

Pure Scanning

Unlike the common plug-in solutions through the TWAIN interface,VueScan does not tie up the photo editor during scanning. You can scan and edit in parallel, which is ideal for batch scanning. The software itself is lean, but the way it works as a RAW editor requires a large amount of memory.

One Scan Is Enough: New Variations Without Rescan

VueScan is designed to scan an image only once. The RAW data of a scan remain in memory until the scan for the next image has begun. It is possible to save the raw data and to make the scan permanently available. Any image manipulation that is possible before the scan can still be performed later on the RAW files – even hardware-based scratch removal. This is because VueScan saves the infrared channel separately and does not apply it to the image. Neither Nikon nor SilverFast can do that.

10.2 Look and Feel

In comparison to the colorful interface of SilverFast, VueScan appears somewhat sparse. Yet, it is structured very logically and is easy to navigate. The main functions are accessible through the tabs Input, Crop, Filter, Color, Output, and Prefs. VueScan maintains ease of use despite an abundance of features. Each menu tab can be switched between More and Less views, in order to display either just the basic functions or all of the available options.

This control of options displayed allows the beginner to slowly get used to the program's extensive functions. The range of options in More mode can sometimes be overwhelming, but there is comprehensive on-line help available that describes each function in detail.

This chapter describes the 8.4.62 version of the software. Since new versions quite often introduce small changes, the screen shots shown in this chapter may differ slightly from your version. The display of functions varies according to the chosen options. Unnecessary switches are hidden, not greyed out. Initially, this can be confusing when you are looking for a particular function and not finding it. The display also depends on the specific scanner model; VueScan only shows options supported by the particular scanner. But all in all the VueScan solution is a good compromise between rich features and an interface which is still quite slim and easy to handle.

▲ The interface of VueScan is simple, yet functional.

Input

Under the Input tab, input options can be found. VueScan not only supports many scanners as input devices but can also open RAW data from digital cameras and scanners. With the right options, it makes no difference for the program whether the data come from a scanner or from a file. (Note: my attempts to import NEF files from an LS-5000 scanner failed – apparently, VueScan's NEF support refers only to Nikon digital cameras.)

The 64-bit RGBI option under Bits per Pixel is especially interesting. With this option, the infrared channel data are saved in addition to the three color channels. The penalty is a 33 percent larger file size compared to a standard RGB scan, but the option allows scratch removal with Infrared Clean afterwards.

Under Scan Resolution, the exact optical resolution of the scanner needs to be set to avoid degrading image quality.

With the Batch Scan option, you can have VueScan generate previews from a folder containing RAW files in one go. Depending on the number of images, generating previews can take a while; but later, during editing, the function allows you to quickly scroll through your collection. Unfortunately, there is no thumbnail view like that in SilverFast or Nikon Scan. To switch between images in a batch scan, you have to move

▼ Scans with multiple passes are possible even for scanners without a multi-sampling feature.

the slide Frame Number or type the frame number directly. The option Number of Samples is available only if the scanner supports multi-sampling (for example, the Nikon Super Coolscan 5000 ED). If your scanner is not multi-pass capable, VueScan can be used to retrofit this function. In this case, VueScan provides the alternative option Number of passes. VueScan scans the image multiple times and assembles it into a single file. This is a genuine feature extension compared to Nikon Scan, and by now SilverFast can do it, too.

Automatically Generating and Saving Preview Images

The switch Preview under the option Auto Scan is handy for automatically generating a preview when the original is inserted. VueScan can save not only the actual scan, but also the preview image. The switch Preview under Auto save will take care of that automatically. This is handy when scanning a lot of material for a quick preselection. Usually, only a few selected images are worth a full scan.

Saving Settings in .ini Files
*The file **VueScan.ini** stores all configuration changes in VueScan. You can store configurations in this and any other discretely named *.ini file. It makes sense to make separate files for different scanning processes, such as slide, negative, or scan from disk. Stored configurations can be loaded on demand with File ▸ Load Options. To restore the factory settings, simply delete VueScan.ini and restart the program.*

Crop

Sometimes, cropping the scanned image can be a little tricky. Especially in batch mode, cropping can lead to black borders. VueScan has a wide range of options to fine-tune each cropping parameter. The switch Auto under Crop Size is particularly useful. The software automatically adjusts the size of the crop. It works well for images with normal exposure, but does not always work with dark images, such as night shots. When cropping the image after the scan, you will always get black borders from the edge of the negative. Although the black edges can be trimmed, they have already caused the exposure to be altered, because automatic exposure treats them as black parts of the image.

The options Buffer and Border are intended to deal with these cropping issues. When enlarging the crop and deliberately scanning with the black border included, the slider Buffer can be used to control the percentage of the image inside the scan frame to be ignored for metering.

Alternatively, you can leave the crop size on Automatic and add a safety margin with the Border function. Now, the automatically selected crop will be scanned with a safety margin, but only the area inside the crop will be used for metering. Unfortunately, in the preview there is no indication of a margin that VueScan may have added (for example, something like a second frame). You can only guess as to what the affected image area is. The combination of Auto Crop and Border is particularly useful in batch mode. If the automatic cropping function has cropped the images too much, you will notice the shifted border.

Filter

VueScan's restoration and correction filters are found under the Filter tab. Unfortunately, these filters are rather meager compared to Nikon Scan and SilverFast. Restore Colors and Restore Fading are not adjustable; they can only be switched on or off. The same applies to the option Sharpen, but this is less of a problem, since sharpening functions should be turned off for scanning film material anyway.

Shifting the Crop Box with the Mouse

By holding down the Shift key, you can use the mouse to move the crop border within the image. Just press ⇧, hold it, and now you can readjust the crop box.

It turns out to be an advantage that VueScan does not apply any of these filters to the RAW data of the image. Unlike Nikon Scan, VueScan generates true RAW data. No corrections are taken into account with RAW data; corrections are typically applied later in processing. The Infrared Clean scratch removal does a fairly good job, but it is not as powerful as Digital ICE. The following example again shows that software has a major impact on the result of the scan. You can use the Color tab to mark scratches that the program detects in red. This is useful for finding the best correction level for the image (Light, Medium, or Heavy).

▼ The red pixels are scratches identified by VueScan.

Color Settings

VueScan's lavish set of tools for color management can be found under the Color tab. ICC profiles for printers, scanners, monitors, and even custom ICC profiles for films can be employed. VueScan also ships with an extensive selection of profiles for common negative films, as well as profiles for Kodachrome and regular slide film.

Though the color management is lavish, VueScan image editing functions are rather poor. The Image menu offers viewing options for various histograms, although viewing is just about all you can do with them. This function is somewhat sparse compared to Nikon Scan and Silver-Fast, but VueScan uses a completely different approach from the other two.

No Post-Processing

VueScan is purely a scanning program; the scanned image is intended to be processed in an image editor after the scan. The image is processed just enough to provide the full image information of the original for later use. For post-processing, VueScan offers some assistance with Pixel Color. If this option is activated, VueScan color codes those image areas that still need some work. The scanned image no longer contains these markings. The function Clipped Black Color marks lost shadow details (in dark blue, compare image next page). The function Clipped White Color marks lost highlights (in green, compare image next page). Presence of both green and blue pixel markings usually indicates that the settings for Black Point and White Point are off and should be corrected either with the sliders or directly in the histogram. VueScan does not utilize a gray point for adjusting mid-tones.

▼ With VueScan, the infrared channel can be saved separately as an image file.

▲ Both black and white points should be pulled outwards. The uncorrected curve is clipped, which leads to lost levels.

▼ Lost shadows are blue, lost highlights are green; black and white points need correcting.

Output Parameters

VueScan offers a rich set of output options; there is no other scanning program currently available which has a comparable functionality. VueScan can even output several image files simultaneously. It is possible to save a scan together with its RAW File and Index File. The index print works just like the corresponding function in conventional photography. VueScan generates a thumbnail from each image and adds it to the central index bitmap. When scanning several films, you should assign different index files to give each film its own index.

The default naming convention for RAW files is raw0001+.tif. The plus sign ensures that VueScan will pick up the count from existing files. Make sure to pick a naming convention that clearly distinguishes RAW files from regular TIFF files. The file extensions are the same for these different files, which can be confusing. VueScan is capable of outputting just about every possible format you can think of, and you can even save a separate image of an image's infrared channel. This aspect of the program's functionality is definitely superior to the Nikon Scan and SilverFast offerings, and it is a nice playground for technicians, too.

Prefs

The basic configuration of VueScan takes place under the Prefs tab. Configurations are saved in the current *.ini file, where configurations for special scans are also stored. A clean separation of special configurations from global settings would be better, but this is not supported in the current edition of VueScan. Nikon Scan has solved the same problem in a better way with its Preferences for global settings. There are various ways to customize the VueScan interface to your specific needs. For instance, with Single tab panel, each tab can be displayed in full-screen mode. This way, you also get a full-screen preview. With External editor, scanned images can be opened in a photo editor of your choice right after scanning. This is convenient for scanning single images.

Batch Scanning and Memory Limits

The Release Memory switch controls whether or not RAW data should remain in memory. For single scans, it makes sense to keep RAW data in memory, because you can access its data quickly. If you change a setting for the image, you do not have to rescan with the Scan button. It actually suffices to do one more Save, which is much faster.

However, in batch mode, this option will dramatically increase the memory requirements. Remember that a single scan of a 35 mm slide can be up to 160 MByte if you are just scanning 4000 dpi. If you are scanning with a Nikon Coolscan 5000 and the slide feeder, this will bring every computer system to its memory limits in a short while. In unattended batch mode, the Release Memory switch should be deactivated, however, because it will use up too much memory if you are opening every file.

Beep When Done

The switch Beep when done is a simple yet ingenious feature. It indicates the end of a scan job with a beep tone, which is very handy for batch scans. You do not have to supervise the scanning process; you can do other things while the system is scanning. The system informs you with the beep when the current scan is finished.

▲ Film profiles included with VueScan.

▼ In parallel with the actual scan, VueScan can also output thumbnails.

VueScan's configurations under the "Prefs" tab.

10.3 RAW Data

VueScan's RAW data processing is a powerful feature. VueScan uses a different approach from Nikon Scan for RAW data processing. VueScan is content with simply writing the RAW data to disk. Unlike with Nikon Scan, it is not possible to edit the RAW data directly. Therefore, VueScan requires a two-step process for editing:

1. Create and save RAW data.
2. Use VueScan to open this RAW data. For VueScan it makes no difference if the data comes directly out of the scanner or from a RAW data file.
3. Generate the final scan, an image file with all filters included.
4. Edit this image with any image editing program. Photoshop would be a good choice.

As far as storage is concerned, this method is more costly than the Nikon solution regarding memory consumption. At least two files exist for each image. Although VueScan RAW files can be opened with a normal photo editor, the results will be unsatisfactory because gamma correction is missing. Only in the preview can the image be evaluated. This does not make it very easy to sort through RAW files. VueScan does not offer a browser view comparable to those provided in Nikon Capture or SilverFast HDR.

Without a doubt, the Nikon solution is much more convenient than VueScan. With the scanning software, the viewer, and the Capture NX2 editor, Nikon has three well-matched components for a smooth RAW workflow. VueScan has only the scanning program, but the integration in third-party applications got better through the last decades.

For a long time, it was only possible to save RAW data in *.tiff format. Besides the fact that you are always in danger of mixing up RAW with standard TIFF data, this format has a major disadvantage. There is no image viewer for this format. Of course you could take nearly every image viewer on the market for it, but the pictures will look shabby. The reason for this is that there is no gamma correction applied.

Keeping Basic Configurations for Different *.ini Files

All configurations of VueScan are stored in *.ini files. To avoid repeating the same configuration steps when using several *.ini files, it is recommended to proceed in two stages: first, create and save a basic configuration with all the settings that are common to all subsequent *.ini files. This includes items such as the ICC profiles of the scanner and the monitor. This configuration can be saved as VueScan.ini, thereby making it the new basic configuration. Next, build additional .ini files for special purposes – such as slide or RAW data scan – which are then loaded as needed.

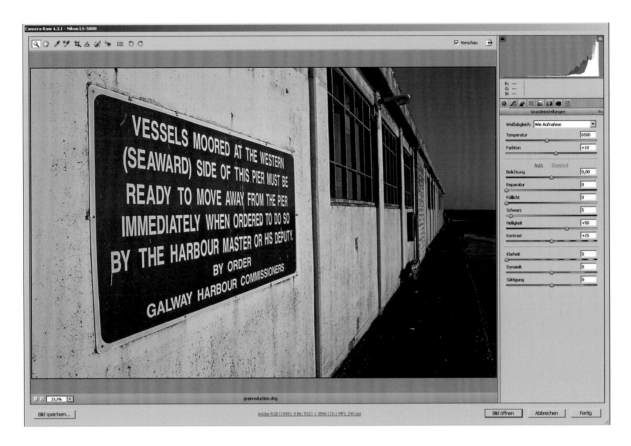

For a proper view, you have to open the RAW data with VueScan first. This process is comparable to a standard scan process, besides the fact that you are actually not scanning but only opening the RAW file stored on the hard disk. In Preview you see the gamma corrected image. If you are handling loads of image files simultaneously, this method is quite annoying. You will miss having a proper browser view like you have in Nikon Capture NX2 or in SilverFast HDR.

But do not worry, VueScan now has a solution for this. It supports Adobe DNG, which is an open RAW data format integrated in the latest versions of VueScan. With DNG, you can conveniently open and sort the pictures with a Viewer like Adobe Bridge. If you want, you can even open the DNG files from Adobe Bridge with Adobe Camera Raw and edit them directly. But be aware that you will then have limited functionality. Adobe Camera Raw does not support scan specific functionality like Infrared Clean. For proper editing, you should still use VueScan. Unfortunately, it is not possible to convert TIFF-RAW-data into DNG-format; in this case, a rescan of your slides is unavoidable.

Overall, VueScan has an interesting approach to RAW data processing. It is the only program that allows you to apply all filters after the scan. One limitation is the quality of those filters. Infrared Clean is not as good as Digital ICE, but this will hopefully change in the future.

▲ VueScan DNG raw scan data can be opened using Adobe Camera Raw. The scan-specific functions are, however, only available in VueScan itself.

10.4 Documentation

The documentation of VueScan leaves a mixed impression. The program does have extensive online help that describes all functions in detail, but there is no search function for its content. Still, you can search each chapter with the browser. VueScan's Help content has been vastly improved compared with earlier versions, and now includes many visual examples instead of just simple text help as in the past. Nevertheless, I still hold to my earlier statement that, while VueScan's Help is adequate for technically-minded users, it is not yet detailed enough for the typical end user.

10.5 Conclusion

VueScan offers genuine RAW data support and sophisticated color management at a comparatively low price. Despite its logical layout, the interface is rather plain and takes some getting used to. Image editing functions are very basic. Unlike Nikon Scan and SilverFast, VueScan is a program intended purely for scanning. In connection with a good photo editor such as Photoshop, it provides an exceptionally rich set of functions. For the technically inclined, VueScan offers a number of capabilities that are not available with Nikon Scan or SilverFast. Among the three products, VueScan provides the most inexpensive way to use the same interface for flatbed and film scanner. VueScan's weakness lies in its filters, which are often not very powerful. Nikon Scan and SilverFast are both superior in this respect. Generally though, VueScan is the most comprehensive of all currently available scan software packages. For technically advanced users, it should be the first choice. No other program is comparably powerful in RAW data processing.

▼ VueScan's online help has come a long way, but is still not exactly user-friendly.

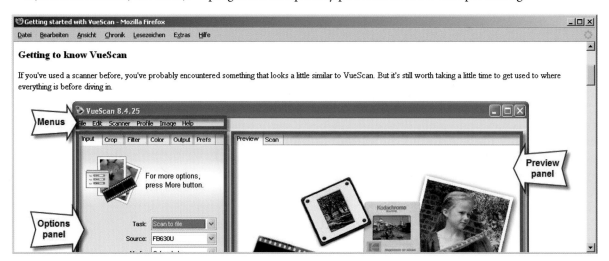

SilverFast

LaserSoft Imaging offers the SilverFast family of programs for scanning and image processing. The basic advantage of the SilverFast family is the consistent user interface used throughout the products. Once you are familiar with the scanning programs SilverFast SE or SilverFast Ai, you will feel right at home with the RAW editor HDR. The extensive image processing functions of SilverFast Ai reach beyond what Nikon Scan and VueScan provide.

11

Contents

SilverFast v6.4.4r3b

▲ In the main window, you can set scan resolution, color depth, filters, and image type.

11.1 SilverFast SE, Ai, Ai Studio and HDR

Several versions of SilverFast are available. The basic version, Silver-Fast SE, is available for download from www.silverfast.com for $49 in the manufacturer's web shop, but it is also bundled for free with many currently available scanners. The advanced version, SilverFast Ai, has an enhanced set of functions and is more suitable for demanding scans than SilverFast SE. Consequently, I will refer mainly to SilverFast Ai in this chapter.

SilverFast supports more than 300 different scanners. Each version is compatible with only one specific scanner model. For compatibility with other scanner models, additional licenses need to be purchased.

SilverFast Pricing: Pay as You Earn

The prices for the various Ai versions depend on the scanner model. SilverFast costs $218 when purchased for a Coolscan V or $398 if purchased for a Coolscan 9000. The Ai Studio versions of the software cost $332 and $512 respectively. The main additional features of these extended versions are AACO and Multisampling. Ai Studio offers further improved functions, such as a 16-bit cloning tool.

With the Ai versions, you can scan images to the proprietary RAW format TIFF-HDR and save them to disk. These files can later be edited with SilverFast HDR. The SilverFast HDR user interface is very similar to the SilverFast Ai interface; there is little difference between the two in usage, whether you access a scanner in Ai or open a RAW file from your hard disk with HDR.

Installing the Scanning Programs

You can install SilverFast SE or Ai either as a stand-alone program or as a Photoshop plug-in. In the stand-alone form, you must go through the SilverFast Launcher to start the program. If you have installed SilverFast for a number of different scanners, you can use the Launcher to conveniently select and start the appropriate version.

11.2 Look and Feel

The SilverFast interface consists of several windows that can be freely arranged on the screen, but the size of the windows can be changed only to a certain extent. The layout is most suitable for the large monitors commonly used for image processing. The software's user interface is rather antiquated and could do with a complete overhaul. The program itself is also sometimes quite slow. You will see the difference when you first work with SilverFast and switch to VueScan later. VueScan reacts much faster that SilverFast.

SilverFast Main Window

In the main window, you can set all the important scan parameters such as resolution, color depth, and image type. For scanning, you need to specify whether you want to scan in normal mode or in batch mode. One special feature of SilverFast not found on Nikon Scan or VueScan is a setting for Image Type. With that setting, you can select preconfigured filter settings for various types of subjects and lighting conditions. This is handy when you first start using the program, since you can quickly explore several variations without needing a lot of background knowledge as to how those settings work.

ScanPilot

The ScanPilot is nothing but a tasklist. It allows you to activate all the options you want to configure before actually scanning. The ScanPilot leads you step-by-step through configuration where you can manually set all options you need. For beginners, this functionality is quite helpful; experienced users are faster without it.

▲ The user interface is divided into several individual windows.

◀ The icons in the ScanPilot.

Error Message with Filmstrip Adapters

Before you start SilverFast with Nikon scanners, the film must already be inserted into the filmstrip adapter. If the adapter is empty, the program will display an error message and will not start.

Prescan

The prescan window shows a preview and a row of buttons for activating or deactivating various scanning options. The prescan window allows you to preview configuration changes. Unfortunately, it does not offer the Natural/Processed tabs that Nikon Scan does for quick comparison, which makes it difficult to keep track of elaborate changes. All changes are updated instantly in the prescan window.

Unlike Nikon Scan and VueScan, SilverFast does not have the option to generate prescans for an entire filmstrip in one go, and then switch freely between images. This can be annoying when working with a scanner with a filmstrip adapter: every time you switch to a new image, you will need a new, time-consuming prescan. The only way to avoid this problem is by scanning and saving images one by one with the JobManager.

Overview

With Overview, you can get a quick idea of what is on the filmstrip. Double-clicking on a thumbnail starts the prescan. Unlike the function in Nikon Scan, the index view in SilverFast can be printed. The print function for thumbnails in VueScan is actually superior, since several

▶ The SilverFast prescan window: There are no Natural/Processed tabs as in Nikon Scan.

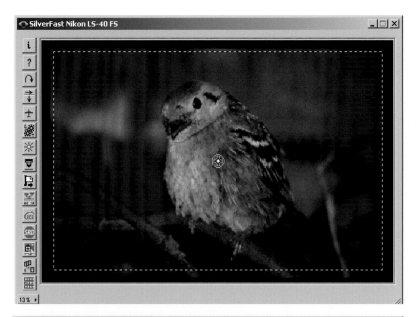

▶ Image preselection in the Overview window: You cannot rotate pictures.

filmstrips can be grouped together in one file for printing. SilverFast's Overview does not allow you to rotate images or to correct the filmstrip offset.

Negafix

As with VueScan, SilverFast comes with a long list of film profiles for the color correction of negative films. When selecting the option Negative under the General tab in the main window, the NegaFix window pops up. SilverFast also allows you to generate and import your own film profiles. You can also fine-tune the profiles using a slider. The option Neg. Direct is available specifically for Nikon scanners; instead of the NegaFix profiles, this option uses the hardware-based negative conversion of the Nikon scanners. This option is useful if NegaFix does not have a profile for your particular film type, or if a profile is available but you are not satisfied with the resulting image quality. SilverFast in Neg.Direct mode, as well as Nikon Scan, both resort to the Nikon MAID libraries for analyzing the film. In general, these librairies produce good results, which can be superior to the standard film profiles. If you are using a Nikon Scanner, you should try them first before using anything else.

▲ NegaFix has individual profiles for all common types of negative film.

Automatic Color Correction During Scan

As of version 6.5, the Negafix tool has an automatic color correction function (called CCR) built in. If the current film profile is no longer accurate due to negatives fading with age, the software can be set to automatically remove the resulting color casts.

Densitometer

The Densitometer displays output values relative to the input value for a small area under the cursor. This is useful for editing colors. If you scan mostly for archiving and do not do much editing, you can just click to close this window.

◀ Densitometer display

Picture Settings

Under Picture Settings, brightness can be adjusted with Exposure, and colors can be made warmer or cooler with White Balance. Mid-tones can be adjusted with Brightness (mid-tones); the gray eyedropper performs the same function. Contrast and Saturation complete the menu. In Nikon Scan, color saturation can be adjusted with the LCH Editor. SilverFast's solution with Picture Settings is less powerful but more userfriendly: histograms are shown in real-time, which sets them apart from the other histograms used for corrections in the program. Real-time histograms give a good view of the level distribution of the scanned image even before the scan is made.

▲ Real-time histogram for output values.

▲ The JobManager is a tool for
unattended scanning.

JobManager for Batch Scans

JobManager allows you to define tasks that the program can execute unattended. Typically, this is used for processing filmstrips. In the SilverFast scanning programs, the job settings are defined for images in the scan queue. In the editor, HDR can handle any image files. For flatbed scanners, it is even possible to create a scan queue with a mix of black and white negatives, color negatives, and color slides in any order. For Nikon scanners, JobManager cannot jump between the modes Negative and Neg.Direct. The user must select one or the other. Once the batch is done, the program sounds a beep, just like VueScan.

JobManager handling in my SilverFast Ai v6.5.5r4 was often quite complicated. Nikon Scan is much easier to use with the Nikon scanner/filmstrip adapter combination, and will enable you to achieve usable results more quickly.

11.3 Set Basic Configuration with Options

General Tab

The General tab in Options allows you to overwrite SilverFast's default setting. This only works, however, if a custom setting has already been defined under Frames ▸ Settings ▸ Save. This is particularly useful if the majority of scans use the same basic setting.

Auto Tab

In order to scan filmstrips at full format, there is an option in SilverFast for Frame Inset. With that option, the exposure metering ignores a certain percentage of the edge of the selected scan area. Unfortunately, there is no preview for this function, and one can only guess the size of the actual area used for metering. With Frame Inset, the black borders included in the scan will not affect the exposure. VueScan offers a similar feature. In Nikon Scan, where this function would also be very helpful, there is no comparable feature.

CMS Tab

The full version of SilverFast Ai includes an IT8 calibration target. Silver-Fast makes it very easy to calibrate the scanner properly. In the prescan window, there is an icon for IT8 Calibration, which lets you generate a scanner profile with a few mouse clicks and apply the profile under Options. The result can then be checked in the prescan window. The IT8 Calibration icon is active only if a scanner profile has been assigned. Unlike in VueScan, calibration is fully automatic in newer versions of SilverFast. SilverFast automatically selects the correct reference file for your IT-8 target and orients the target automatically, too.

Special Tab

In SilverFast, the analog gain is hidden in Options under the tab Special and is accessed under the term Lamp Brightness. In Nikon Scan and VueScan, the best setting must be found by trial and error. This can be tedious since each new setting requires a new preview scan. SilverFast has a better solution in the form of a real-time histogram for the estimated effect on the image. This is a unique feature which is quite handy in daily use. For some unknown reason, this Lamp Brightness feature is placed under Options, where it globally affects all scans. This makes little sense. A classic use for the analog gain control is for correcting the occasional underexposed or overexposed image; it would be better, then, to place this control at the frame level where it can be applied to each scan individually.

Speeding up Scans

If you use the preview image mainly for setting the crop and the black/white points, then you do not need a high-resolution color preview. Instead, you can use the switches Prescan Draft and Prescan Monochrome, and reduce the time for scanning the preview image. Avoid activating Scan Draft in all circumstances. This feature severely reduces the quality of the full scan, which is something you should not accept when scanning your slides for archiving purposes.

Aging targets
Calibration targets should be stored in a lightproof environment, but they are subject to deterioration with age even if stored carefully. If you need to replace your targets, you can either go to the SilverFast web shop or visit www.coloraid.de.

◀ Settings for color management.

◀ Analog gain in SilverFast with real-time preview.

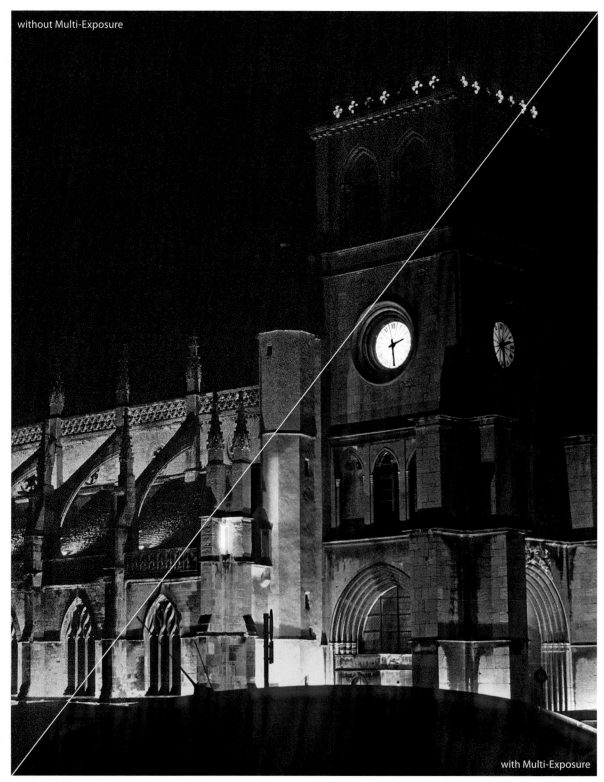

without Multi-Exposure

with Multi-Exposure

The Multi-Exposure function can be used to increase your scanner's dynamic range, but at the price of doubling the required scan time.

11.4 Image Processing

Image processing in SilverFast is extensive and greatly exceeds the abilities of VueScan and Nikon Scan, let alone the common TWAIN drivers. SilverFast offers many functions of a full-scale image editor. Therefore, the program is well suited for image processing during scanning.

Selective Color Correction (SCC) and the Clone Tool are two good examples of advanced image processing functions. SilverFast does not reach the depth of Photoshop, of course. Whether or not the enhanced functionality of SilverFast is needed depends on which workflow you chose. If you post-process in Photoshop anyway, then VueScan and Nikon Scan will be sufficient. For space reasons, I will elaborate only on color correction here. A complete overview of all options and functions can be found in the excellent SilverFast PDF manual.

SCC is similar to the Nikon LCH Editor, but it is easier to use. Furthermore, in SilverFast, corrections can be restricted to certain image areas with the use of masks, which is not possible in Nikon Scan. In the first step, take the eyedropper and pick the color to be changed. In the example on the previous page, it is the green of the excavator. To boost the green, draw a line from the center of the disk toward the green point on the outer circle. Next, select the color of the blue sky and boost it the same way.

With SCC you can boost selected colors, change their characteristics, or even replace them completely. The easiest way to change color characteristics is to assign different output values to the colors in the table. The densitometer comes in handy here; it displays all color changes in real time.

11.5 RAW Processing with SilverFast HDR

SilverFast HDR supports not only RAW files created with SilverFast Ai, but also various RAW files from digital cameras.

Creating RAW Files

With Ai you can save RAW files in the formats 48-bit HDR Color and 16-bit Grayscale. The latter format saves storage space for black and white images, but using it can cause problems later, since not all image editors fully support grayscale mode.

Unlike VueScan, SilverFast does not save the infrared channel separately in HDR. Doing an ICE correction later is not possible. Still, all other SilverFast filters can be applied to the RAW data afterwards, such as GANE for grain reduction.

As with VueScan, SilverFast does not save changes directly in the RAW data. Both programs write a second file and leave the RAW file to serve as a digital negative. Nikon's NEF concept is more user-friendly, since everything is contained in one file.

▼ RAW files are saved in HDR TIFF format.

Preselection with Virtual Light Table (VLT)

SilverFast VLT is the file browser and image management software from LaserSoft Imaging. The virtual light table helps sort and organize RAW files for processing in HDR. In addition to a number of RAW formats from cameras, HDR can also handle the SilverFast Ai TIFF-HDR format. It does not recognize Nikon scanner NEF files, but it does recognize NEF files that were produced using digital cameras. The program also supports the following camera formats: CRW (Canon), CS (Sinar), DCS (Kodak), MRW (Minolta), ORF (Olympus), RAF (Fuji), RAW (Leica, Panasonic), SRF (Sony), and X3F (Sigma). TIFF RAW files from VueScan cannot be processed since they have no gamma correction.

VLT Automatic Closedown

With SilverFast VLT, files to be opened in HDR can be selected by double-clicking the files on the light table. As soon as HDR opens a file, VLT closes, which initially can be confusing. The light bulb in the JobManager or in the prescan window allows the user to switch between HDR and VLT.

Automated Batch Processing

The link between HDR and VLT is the JobManager. JobManager is always required for batch processing. Files are processed by selecting them with the mouse and dragging them into the JobManager window. When adding a file, the Options window opens and prompts for Settings and Picture Type. Only then will the image be added. Clicking on the

---

RAW Conversion of an HDR Negative: Also Possible in Photoshop
Unlike the gamma-uncorrected VueScan RAW files, HDR files can be easily converted in Photoshop. First, reverse the colors with Image ▸ Adjustments ▸ Invert. This action provides a positive, but still requires serious color correction. The easiest way to correct color is with Image ▸ Adjustments ▸ Auto Levels.
It is helpful to scan a bit larger and include the frame border. This portion of the negative is a perfect reference for the color black (for example, to be selected using the black eyedropper).

Processing icon opens the image in HDR. Only one image may be open at a time. Overall, the way VLT is handled in connection with scanner RAW files is not very intuitive. Only VLT or HDR, but not both, can be open at one time, which is a problem since it is often necessary to jump between them during image processing. Especially with larger monitors, it would helpful to keep VLT open next to HDR.

▲ The user interface of HDR is mostly identical to SilverFast Ai.

Scans from Disk with HDR

Working with HDR differs only slightly from working with SilverFast Ai. One difference is the absence of ICE, because in the SilverFast RAW format the infrared channel is not stored separately. ICE must be invoked during the scan. However, batch processing of filmstrips is much more convenient in HDR than in Ai. With the JobManager, you can easily click your way through a long list of files. Previews will update to the selected image within seconds.

Using HDR makes sense if you split the scanning process into two steps: first, scan RAW files with only minimal adjustments. Then, add them to the JobManager with VLT. In JobManager, set the parameters for each image and save them. After all images have been prepared in this fashion, click on Start to begin the batch processing. Without any further user interaction, all images will then be written to the predetermined file location. As you can see, SilverFast's HDR is an interesting approach to RAW data processing but has some limitations in functionality.

▼ The EXIF data contains information about the scanner and the scan.

Show brightest or darkest area within current frame. Adds measurement point with <Shift-click>

▲ SilverFast conveniently displays the white point to help with white balance. In Nikon Scan, you have to find the white point yourself.

11.6 Documentation

None of the competitors has documentation as detailed and extensive as SilverFast. PDF files and video clips lead you through all the sophisticated software features.

The main SilverFast document is a 494-page manual in PDF format. The extent and quality of the PDFs, which can be downloaded free of charge, are actually better than the official SilverFast guidebook, which costs $23. For many functions, there are clear examples with images. In the program itself, most buttons display context-sensitive help when the curser hovers over them.

11.7 Conclusion

Just like VueScan, SilverFast is a program that supports many scanners and often gets more out of slides or negatives than do the TWAIN drivers included with most scanners. LaserSoft's restrictive licensing policy can be quite costly, especially if you own several different scanners. Also, it is not quite clear why the Ai version for the Coolscan 9000 is priced so much higher than the same version for the Coolscan V.

The main feature differentiating SilverFast Ai from Nikon Scan and VueScan is its extended set of image editing functions. The high cost of SilverFast can be justified only if you actually do all image processing during scanning and not in Photoshop afterwards. However, doing most of the image processing in Photoshop, or any other good editor, is actually a far better approach.

▲ 494 pages: the SilverFast-Manual in PDF-format has almost 30 Megabyte. It is available in different languages.

Total Integration

One nice thing about SilverFast is that the scanning program, image editor, and image management functions are all very well integrated. However, the user interface, with all its scattered windows, is a real weak point. Working with SilverFast is quite clumsy without a large monitor. The workflow for batch scanning with the filmstrip adapter also shows clear weaknesses; in particular, there is no cache for prescans.

With its unique combination of IT-8 calibration and Digital ICE functionalities, SilverFast is definitely an interesting alternative for owners of Nikon scanners. Nikon Scan has no calibration feature and VueScan scratch correction is no match for ICE. On the other hand, the many automatic functions in the program are helpful for the beginner. Even advanced users will find the excellent product documentation very useful to help them exploit the full spectrum of SilverFast. The program's documentation is rounded out with a collection of video tutorials.

▶ With Nikon Scan, the image has a
slight color cast

▶ The IT8-calibrated SilverFast shows
more neutral colors.

▶ In both cases, the image should be
fine-tuned for best results.

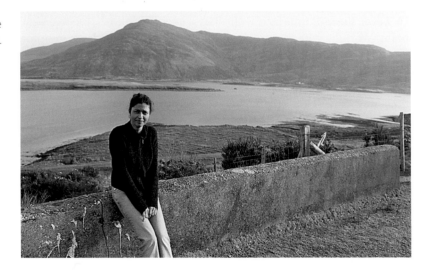

Scan Workflows

The workflows introduced here have been proven in actual use with several thousand slides and negatives. To speed up processing, the editing functions of the scanning software are rarely used during scanning. The first priority is to capture the originals in full quality in a reasonable amount of time and get them onto the hard disk in the form of RAW data. This is always the primary goal when trying to efficiently process a large quantity of material. There are many opportunities for creative processing with an image editor after the scan has been completed - and they do not require spending time waiting for the scanner, either.

12

Contents

12.1 Vital Corrections before Scanning

Raw data formats allow you to retroactively apply most filter and configuration parameters to a scan file, but hardware settings (such as focus or Analog Gain) cannot be subsequently applied.

For technical reasons, the following settings must always be made prior to scanning. Any options not mentioned here can be taken care of later, either in the scan software or during image editing with Photoshop.

1. Set Scan Resolution
To archive originals in good quality, it is recommended that they be scanned at the optical resolution of the scanner. It is no problem to down-sample the scan later. It does not work the other way around, though, since up-sampling only increases file size without gaining image information.

2. Select Sufficiently Large Crop
Any image that has been cropped too tightly cannot be magically restored later. However, cropping too generously affects the exposure with Nikon Scan, because the black film border will be taken into account to determine exposure. It is different for VueScan and SilverFast. With these, the border can be configured freely, yet can be ignored for exposure metering. At any rate, it is always important to capture the entire image.

3. Set Focus
Only a correctly focused original can generate a sharply scanned image. For flat film material, autofocus works quite reliably in most cases. Warped originals require a manual focusing point. It is best to flatten warped originals in a stretching frame prior to scanning. Only then can the scanner capture the entire area sharply.

4. Adjust Analog Gain
Adjusting the scanner's light source can compensate (within limits) for underexposure and overexposure in the original. This physical effect cannot be reproduced by subsequent correction in an image editor.

5. Perform Scratch Removal with ICE or FARE
Neither Nikon Scan nor SilverFast can perform ICE correction after scanning. VueScan, on the other hand, can store the infrared channel in the RAW data and can later perform the function Infrared Clean.

6. Set Level Adjustments: Shifting Black and White Points
Spreading out the level spectrum before the scan optimizes the level distribution in the image. The correction can also be performed after the scan, but with a somewhat inferior result. The next section compares the effectiveness of level adjustments before and after scanning.

Level Adjustment: Before or After the Scan?

In many cases the original does not fully utilize the entire tonal range from 1 to 255. By setting the black and white points, or by manually adjusting the curves, the level spectrum can be spread out either before or after scanning. Applying a correction before the scan is always best. In that case, the scanner does not waste data on levels that do not even exist in the image. Correcting before the scan distributes the levels evenly.

Drop-outs with 8-Bit

As can be seen from the example, stretching the levels after the scan deteriorates the image. The subsequent correction caused massive drop-outs in the levels. The correction before the scan is significantly better, but there are still undesirable spikes in the 8-bit file.

Bigger Reserves with 12-Bit

The second example was scanned with a 12-bit color depth. The subsequent correction generates fewer drop-outs in the levels, which demonstrates that the 12-bit format has bigger reserves. But, even with 12-bit being used, the result of the correction before the scan is a little better than the correction afterwards. The greater the color depth of the scan, the better. The 16-bit scans of modern scanners provide plenty of margin for later corrections. Still, if possible, the black and white points should be set prior to the scan to fully utilize the scanner.

Top row (8-bit scan): on the left, the original condition; in the middle, corrected after the scan; and on the right, corrected before the scan.

Lower row: The same sequence as the previous example, but scanned with 12-bit. Here the curve shows hardly any drop-outs.

12.2 Nikon Scan Workflow

Scanning Slides Individually and with a Slide Feeder

1. **Change and Save the Default Settings**
   Set ICE to Normal and the color depth to the maximum supported (12-, 14-, or 16-bit). Activate ROC and GEM only if needed. In conjunction with ICE, the suppression of hard edges with GEM is already evident at level 1. Do not use DEE.

2. **Create Preview**
   Under Preferences and Automatic Actions, simply inserting the slide triggers the preview.

3. **Adjust Preview**
   Position and adjust the crop, if needed. Set black and white points. Alternatively, use the sliders for Curves. For warped originals, manually set the focus point on a key detail in the image. Set Analog Gain only for severe underexposure or overexposure.

4. **Scan**
   Save the scan as NEF or TIFF. Even with black and white negatives do not use the option Grayscale-NEF to avoid compatibility problems in post procesing.

5. **Post-Process with Capture NX2 or Photoshop**
   Finalize cropping, set gray point, and rotate images. Optimize brightness and contrast with Curves. Eliminate color casts with the gray eyedropper. Balance shadows and highlights with either D-Lighting or Shadow/Highlight.

Special Note for Batch Scanning with Slide Feeder

Skip steps 2 and 3 when batch scanning. After saving default settings, Slide Feeder Scan will let the scanner run for up to 99 frames. Level adjustments can be performed afterwards with Nikon Capture NX2 or Photoshop.

Nikon Scan:
Batch Scanning of Negatives with Filmstrip Adapter

1. Change and Save the Default Settings
Set ICE to Normal and the color depth to the maximum supported (12, 14, or 16-bit). Activate ROC and GEM only if needed. In conjunction with ICE, the suppression of hard edges with GEM is already evident at level 1. Do not use DEE.

2. Generate Thumbnails, Correct Filmstrip Offset
Automatically start the index scan with Automatic Actions under Preferences. Manually correct the Filmstrip Offset, if needed. Preselect images and manually start the preview for only the chosen images.

3. Assign User-Defined Settings and Adjust Preview
Mark previews on the index and assign settings. Set the suitable crop for each image by hand. Set black and white points. For warped originals, manually set the focus point on a key detail in the image. Set Analog Gain only for severe underexposure or overexposure.

4. Scan
Save the scan as NEF or TIFF. Even with black and white negatives, do not use the option Grayscale-NEF to avoid compatibility problems in post-procesing.

5. Post-Processing with Capture NX2 or Photoshop
Finalize cropping, set gray point, and rotate images. Optimize brightness and contrast with Curves. Eliminate color casts with the gray eyedropper. Balance shadows and highlights with either D-Lighting or Shadow/Highlight.

Don't Assign Settings for the Batch Scan before Preview!

Only when settings are assigned to the individual images after the preview can they be completely implemented for the scan afterwards. If the settings are already assigned to the thumbnails, the images will reset to the default after the preview. According to the manual, this strange behavior is not a bug, but rather a feature of Nikon Scan.

12.3 SilverFast Workflow

Digitizing Black and White Negatives

1. **Generate Overview Scan**
   Select the desired image from the overview scan; the preview will start automatically.

2. **Adjust PreScan**
   Set the crop and, if needed, rotate or mirror the image. Set the black and white points with either Histogram or Auto Adjust. For warped originals, manually set the focus point on a key detail in the image.

3. **Remove Scratches with SRD**
   SRD requires a second pre-scan. For that, the output resolution must be set to the final value. Set SRD to Expert Mode and configure Dust and Scratch Removal, as well as Defect Size. Check the preview window for defects (marked in red).

5. **Scan**
   Start the scan in 16-bit grayscale.

6. **Post-Process with Photoshop**
   Fine-adjust Midtones (gray point) with Levels, tune brightness and contrast with Curves. Balance shadows and highlights with Shadow/ Highlight.

No Batch Scanning with SRD

Manual scratch removal for each individual image cannot be automated. Therefore, it is not possible to automate batch scans for black and white negatives. Currently, SRD cannot even be used with the JobManager. Alternatively, all images can first be scanned in RAW format, and the RAW data can then be edited in SilverFast HDR. This way, at least for step 2, there is no wait for the scanner.

12.4 VueScan Workflow

Scanning Single Slides

1. **Set Defaults for RAW Data**
   Set Crop Size to maximum, deactivate all filters, and for Output select the option RAW File in 64-bit RGBI mode. Save settings with Save Options.

2. **Generate and Adjust Preview**
   Through Auto Scan in Preview, a preview will automatically be generated when a slide is inserted. Roughly set Black Point and White Point. For warped originals, manually set the focus point on a key detail in the image. In case of severe underexposure or overexposure, set analog gain with Red/Green/Blue Analog Gain.

3. **Scan and Save**
   VueScan automatically saves to hard disk if Auto Save is set to Scan.

4. **Load RAW Data**
   Switch Source to File and open RAW files from disk. In Batch Mode, with sufficient memory, an entire folder can be opened and placed in a cache. Film profiles should only be set for Kodak; otherwise, use Generic.

5. **Process and Save the Scans from the Hard Disk**
   Treat RAW data like a scan from film: Set Infrared Clean and other required filters, such as Grain Reduction. The Preview window provides feedback. Cropping is most convenient with Crop Size: Auto. Keep the option Auto Rotate active. Save as 48-bit RGB. Save the defaults to simplify switching between modes.

6. **Post-Process with Photoshop**
   Fine-tune Black Point, White Point, and Midtones (gray point) with Levels. Tune brightness and contrast with Curves. Rotate and mirror, if needed, with Rotate Canvas. Balance shadows and highlights with Shadow/Highlight as needed.

VueScan:
Scanning Negative Strips with the Filmstrip Adapter

1. Change and Save Defaults for RAW Data

Set Crop Size to maximum, deactivate all filters, and for Output select the option RAW File in 64-bit RGBI mode. Save settings with Save Options. Leave Batch Scan set to Off.

2. Generate Preview and Correct Filmstrip Offset

Manually generate a Preview for the second image, and manually correct the filmstrip offset. It makes no sense to generate all previews and set the black and white points at this stage, since the current setting always overwrites the values of all the other images. With VueScan, the same default settings apply to the entire filmstrip.

3. Scan and Save

Change Batch Scan to All. Use Scan to start the batch scan. VueScan automatically saves to hard disk if Auto Save is set to Scan.

4. Open RAW Files

Set Source to File and open RAW files from the hard disk. Select a suitable film profile.

5. Process and Save the Scans from the Hard Disk

Working with scans from the hard disk is fast. Therefore, process and save each image individually for best quality. To do that, set Batch Scan to Off. Set the black and white points. Preferably use manual crop. Set Infrared Clean and other required filters, such as Grain Reduction. The Preview window provides feedback. Save as 48-bit RGB. Save the defaults to simplify switching between modes. Batch process the scan.

6. Post-Process with Photoshop

Fine-tune midtones (gray point) with Levels. Tune brightness and contrast with Curves. Rotate and mirror, if needed, with Rotate Canvas. Balance shadows and highlights with Shadow/Highlight.

Nikon Capture NX 2

Nikon's Capture NX 2 photo editing software is a part of the Nikon image processing software system, which consists of Nikon Scan, the Nikon View NX photo viewing program, and Capture NX 2. Although Nikon Scan and Nikon View are available free of charge, Capture NX 2 costs $149.95 for a full license. Capture has no real competition in the field of NEF file editing, and all write commands are non-destructive, allowing you to return to the original image state at any time. This feature is generally not supported by any other conventional image processing program and is Capture's greatest strength. When it comes to processing JPEG and TIFF files, Photoshop & co. are much more powerful tools. Although Capture NX 2 was originally designed for use with digital cameras, most of its functions can be equally well applied to RAW files.

13

Contents

13.1 RAW Data Processing Using Capture NX 2

Differences Between Capture and Photoshop & Co.

▲ Capture NX 2 requires the Microsoft .NET 2.0 Framework to run correctly .

First and foremost: Capture NX 2 was developed exclusively for use with Nikon digital cameras*, and the fact that it is well suited to processing NEF scan files is simply a happy coincidence. Capture's full spectrum of functions is not supported by NEF scan files, but Capture NX 2 is nevertheless currently the best RAW scan editor on the market. Capture has all the major tools you need to give your scanned images their final polish, and includes some RAW processing functions that even Photoshop doesn't offer.

Photoshop Cannot Save NEF Natively

Nikon Capture and Capture NX/NX 2 do not officially support NEF scan files, so if you decide to process your scans this way, you do so at your own risk. I have been using Capture since 2004, and the current (at the time of writing) 2.0.0 version has so far caused me no problems. This is, of course, no guarantee that this will be the case for your system, or for future releases of the software.

Thanks to the proprietary Adobe Camera Raw module, Photoshop can open and process NEF files, but it is still not capable of saving changes to the native NEF format. Saving changes necessitates the creation of a new file—in TIFF format, for example. This uses additional storage space, and extra file versions make preserving an overview of your work more difficult. Capture NX 2 saves changes to NEF files in a separate configuration file that is automatically saved alongside the actual image data. The advantage of this system is that you never run the risk of ruining an original image with substandard image processing.

Photoshop utilizes a completely different approach. Unless you create a second copy of your image, it resamples changes directly into the original image file. Photoshop is also capable of saving multiple image layers, but not all image viewers are capable of displaying them. The

Version tool in Capture NX is a simple and powerful tool for handling different versions of the same NEF image, and it ensures that the original image remains unchanged. Any additional versions of your image are saved separately in the same image file, and they can be opened and viewed at will. In Photoshop, a single botched unsharp mask command can ruin the original image file forever, whereas in Capture NX, you can restore your image to its original state at any time with a single click.

Not Developed for Scanning Purposes

So much for the basic differences between Nikon and Adobe RAW file processing. Adobe Camera Raw works perfectly well, but—as the name suggests—only with RAW files created by digital cameras. The module simply does not work with scan files. Capture NX 2, on the other hand, can open and process NEF scan files, and provides you with all the basic functions you would expect to find in an image processing program: gradation curves, tone value correction, color correction, cropping etc. Capture's limitations are more evident when it comes to the creative aspects of image processing.

Photoshop CS3's Clone Stamp, Healing Brush, and Spot Healing Brush tools provide a whole range of ways to remove dust and scratch marks from your images. For a long time, Nikon offered no such tools as part of the Capture package, but now includes the useful Selection Brush in the NX 2 version, as well as the Photo Effects, Auto Red-Eye, and Straighten tools, along with various other filters. Together, these give Capture NX 2 a more than adequate range of image processing functionality.

Better than Adobe?

Revolutionary features, such as Capture's U Point technology, have actually given Nikon a lead over its old rival Adobe in some aspects of image processing. Control-Points make the use of complicated masking techniques obsolete when it comes to applying color corrections to an image. Nevertheless, it is probably still best to combine programs to make the best of your images.

Capture NX 2 is the best tool available for processing and archiving RAW files, but is still not equal to the task of effective, creative image processing. All common image formats can be opened directly from Capture in an external editor (such as Photoshop) without having to save them first, making it easy to open NEF scans in Photoshop.

The ideal workflow thus consists of making basic adjustments to your scan using Capture NX 2, then using Photoshop where the Nikon program reaches its creative limits.

▲ Using Capture NX 2, you can return to the original, unchanged image at any time with a single click.

13.2 The User Interface

Capture's .NET-based user interface is not very clear and not particularly intuitive, but it can nevertheless be mastered in a relatively short space of time. Experienced Photoshop users switching to Capture will have to get used to a completely different approach to image handling and an unfamiliar user interface. But with new features like U Point Technology, Capture NX 2 is definitely worth a try.

New Browser

Experienced users of the old Nikon Capture program will sorely miss the lightbox tool, which was retired with the introduction of Capture NX. This feature displayed all images contained in the current folder as thumbnails, making image selection quick and easy. The new browser's default behavior is a little irritating; it collapses the entire browser into the application's sidebar when you select and open an image. This behavior is acceptable for use on smaller monitors, as it saves space on the desktop, but users of larger monitors or multi-monitor desktops quickly find it annoying.

To work around the problem, you simply need to undock the browser window from the sidebar by clicking the small triangle icon in the browser's title bar. The browser is not particularly suitable for simple image viewing, as it is fairly slow, especially when viewing large numbers of files simultaneously. Other programs, such as Nikon View NX, are better suited to the job and have much better caching functionality, making viewing a faster, smoother task.

▼ The default setting collapses Capture NX 2's browser into the sidebar when an image is opened.

Dynamic Palettes

Capture NX 2's image processing tools are organized into Palettes, which function differently from those you might be familiar with in Nikon Scan. The File Directory, Browser, IPTC Information, Bird's Eye, Photo Info, and Edit List toolsets each have their own palette. The most important of these for daily use is the Edit List palette, which differs from the one in Nikon Scan in that certain functions (such as LCH) are no longer integrated into the palette itself. If you open the LCH tool using the menu bar, an entry is automatically generated in the Edit List. Every additional editing step then produces its own entry in the list, making it simple to reconstruct exactly which changes you have applied to your image and in what order.

Changes are Fully Reversible

The only disadvantage of this system is that when you apply large numbers of changes to an image, the list takes up a significant amount of space on the desktop. The major advantage is that you can undo each change selectively with a single click on the appropriate step, making it very easy to see what effects you have achieved with the changes you have applied.

Implementing a similar feature would benefit many other image processing programs (including Photoshop), and is a great help to image processing beginners. Each version of the image can be saved under its own name, and you can return to the original version of the image at any time with a single click.

All palettes are docked to the application's sidebars by default, but can be scaled and placed however you like on the desktop. If you should discover that you can no longer find a palette, you can use the Window menu to recall each palette manually.

▼ The Edit List allows you to view and undo each image processing step individually.

◀ Capture NX 2's interface is not very suitable for smaller monitors, and you really need a vertical resolution of at least 1200 pixels to accommodate the application's large tool palettes.

13.3 Important Features

This section deals with some of Capture NX 2's features that go beyond the functionality of Nikon Scan software. These include a number of noteworthy features that are useful not only for RAW image processing.

The Crop Tool

The Crop tool accessed via the toolbar, can be used to crop images manually or to preset sizes and proportions. This is, in itself, not particularly spectacular, as just about every freeware image browser these days includes a working crop tool.

What makes Capture NX 2's crop tool special is that, instead of simply deleting the cropped areas, it lays a virtual frame over the original image and preserves the data from the cropped portions of the frame. This means that you can view the cropped image in the browser and revert to the original, uncropped image at any time. This feature can, however, only be applied to NEF-format images. Cropped JPEG or TIFF images however, stay cropped once the command has been applied and this can not be reversed.

This feature has proved its worth in Capture and Capture NX. It doesn't yet work in my pre-release 2.0.0 version of Capture NX 2, but I am assuming that this is simply due to a bug that will be resolved by the time this book is published.

▼ NX 2's comfortable batch processing is suitable for handling even large numbers of images.

D-Lighting : Successor to Digital DEE

Nikon successfully implemented user-friendly, selective correction of highlights and shadows with the introduction of the Digital DEE feature in Nikon Scan 4.0. The disadvantage of this otherwise great feature is that you have to decide before scanning whether you want to apply it, as the corrections it makes are sampled directly into the RAW image data during scanning and cannot be undone. This approach can be very time-consuming, as you may need to make several scans on the way to achieve optimum results.

Opening Images using Drag-and-Drop

You can open images in Capture using drag-and-drop simply by dragging file names from Windows Explorer to the application icon.

Digital DEE has been implemented since Capture 4.1, but it was initially only usable for files created using digital cameras. As of version 4.3, the tool has been renamed D-Lighting and can now be applied to NEF scans too. Functionally, the tool has hardly changed, although the Threshold slider has disappeared and a Color Boost slider has been added. This allows you to make the color adjustments necessary, for example, after a shadow correction has been applied.

This improved integration of highlight and shadow correction into the RAW data processing workflow is a great improvement, and makes it possible to try out different effects on a finished scan without having to rescan your original. If you have a copy of Capture NX 2, you can simply ignore digital DEE while scanning.

Lost Highlights and Shadows

Lost Shadows are the portions of an image where no detail is visible, even if image data is present. Lost Highlights occur where image detail is simply too bright to be recognizable. Such phenomena only become critical if you want to save your scan to a non-NEF format (JPEG, for example). In this case, the lost shadows and highlights will be lost forever. In order to keep this from happening, you can view the potential lost shadows

▼ Crops of NEF images made using Capture simply lay a virtual frame over the original image, thereby avoiding the irretrievable loss of image data while cropping.

and highlights via the corresponding commands in the View menu. You can then make tone value corrections before saving your image.

Correcting Slanted Horizons

A useful feature since version 4.3 is the Straighten tool. This feature comes into its own if the camera was not held level during shooting, or if the slide was not precisely aligned during scanning. In Capture NX 2, the Straighten tool can be found in the toolbar. All you need to do is drag a line through the image using the mouse, you can also adjust the horizon by manually entering a degree value. The 10° correction limit in Capture has been removed, it can now correct any degree.

Before and After

If you have been working on an image for a while, it is easy to lose sight of whether your changes are actually improving things. Capture NX 2's Compare with Original tool (found in the View menu's Compare command) displays your original and the current image in a single window.

▾ As you can see from the position of the horizon, this photo was not leveled out when taken. With Nikon Capture NX 2, you can correct any degree of slant.

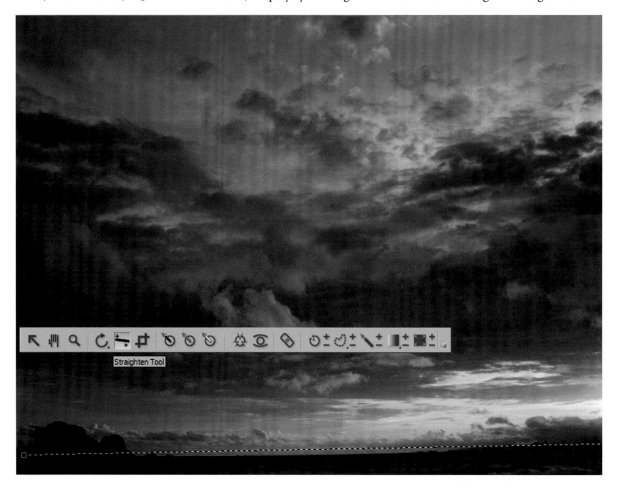

Selection Tools and Other Similar Functions

Well-known Photoshop tools such as Lasso, Polygon Lasso, Rectangle Marquee and Oval Marquee are also available in Capture NX 2, which is unusual for a RAW editor. The application of these tools is, however, somewhat limited, as they are only good for marking portions of an image which are to be processed using Capture NX 2's own tools. Even the simplest processing tools from other manufacturers' image editors (such as cut, copy, or paste) are not supported by Capture.

In addition to the selection tools, Capture includes a Fill/Remove tool that can be used to apply or remove the current effect either for the currently selected area or for the entire image. The Selection Brush is another interesting tool that also allows selective removal of applied effects within the area marked by the brush.

Wedding Photos in RAW format

A classic application of the Selection Brush is the apparently obligatory soft-focus effect applied around flower bouquets in wedding photos. The effect in question involves the flowers at the center of the image being in sharp focus while the entire background fades into romantic softness. Capture NX 2 can now achieve this kind of effect at a RAW data level. All you need to do is apply a Gaussian Blur to the whole image, and then selectively remove the soft-focus effect this produces using the Selection Brush.

Testing Capture NX2
A test version of Capture NX 2 is available free of charge from the Nikon download portal, and can be installed alongside the older Capture NX version without causing any conflicts. The original Capture, however, is not compatible with newer versions of the software, and should be uninstalled before you install any NX variants.

▼ Capture NX 2 can compare the processed version of an image with the original without having to save changes first.

Control Points

Nikon's Control Points are not directly comparable with other, more conventional selection tools like Lasso and Ward from Photoshop. Instead the Control Points are based on the proprietary U Point Technology. Use of Control Points is an easy two-step process. The first step is to mark image areas with one or more Control Points, and the second step is to select the effect, filter, or other change to be applied to the selected areas. The process is much quicker and simpler than the time-consuming masking systems used by other image processing programs. The results of Control Point-based corrections are enough to convince even the critical minds.

The Auto Retouch Brush

Although Photoshop has provided a whole range of stamp-type retouching tools for a number of years, Capture NX 2 is the first Capture version to include a similar general tool. With Auto Retouch Brush you can remove dust spots, blemishes, or other distracting details now even at RAW level. The tool is relatively simple to apply and works pretty well automatically.

Limited Brush Configuration

Scan File Metadata

The metadata recorded with NEF scan files contain important information regarding the scanner model used and the settings (such as the film profile) that were active during the scan. The metadata also reveal whether ROC, GEM, ICE, or Analog Gain were activated during the scan, and if so, the levels they were set to.

When using Auto Retouch Brush, the only adjustment you can make is to the diameter of the brush. Tests have shown that the tool is reliable if you want to remove objects from evenly colored backgrounds, such as an overhead cable in front of a blue sky. However, if the background has finer textures of its own, you can often end up with unwanted artifacts in your final image (in this respect too, the tool is directly comparable to Adobe's Spot Healing Brush).

In spite of its limitations, the Auto Retouch Brush is important and can be useful in many different situations. The tool doesn't require the time-consuming preparatory steps (new levels, safety copy, etc.) that Adobe's tool necessitates, and you can remove all its effects from your NEF files at any time with a simple click.

No Downward Compatibility

NEF files that have been altered using the Auto Retouch Brush can be opened in the old Capture NX, but the effects of the tool itself are not visible in the earlier software version's image browser. At least this does not spoil downward compatibility; you don't have to worry about that when applying this correction.

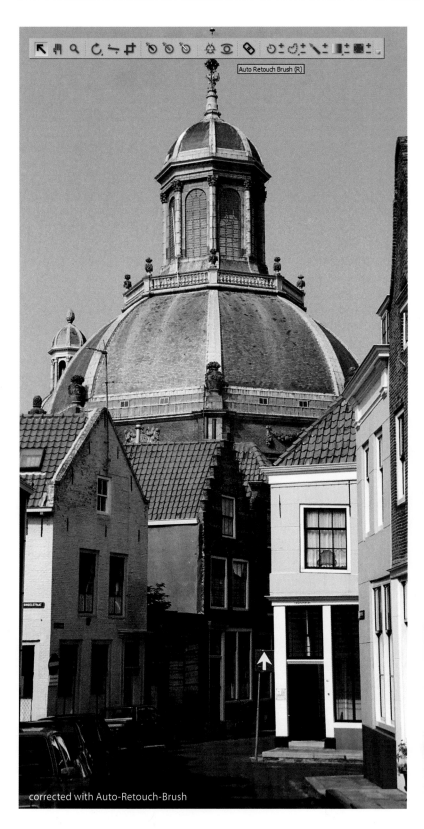

◀ The Auto Retouch Brush works similar to the Spot Healing Brush in Photoshop.

▼ This uncorrected source image needs some retouching, as the lamp and some dark-spotted artifacts should be removed.

corrected with Auto-Retouch-Brush

without correction

Correcting Red-eye Effects in NEF Files

Red-eye is an especially common effect in flash portraits taken with compact digital cameras, due to the very close proximity of lens and flash. The light from the flash is reflected directly from the blood vessels in the subject's retina into the camera's lens, producing the well-known (and unwanted) effect.

Red-eye correction is, these days, a standard feature in most image processing programs, and is even included in freeware programs such as IrfanView and Nikon View NX. Scanner users with a preference for RAW files have so far not been able to enjoy the same freedom, and removing red-eye from NEF files has always involved editing files in a separate program, thus making it impossible to save treated images back to NEF format.

This situation has changed, and Capture NX and NX 2 have continually improved their built-in anti-red-eye functionality. The latest version of the tool lets the user set a manual Red-Eye Control Point, rather than relying on automatic red-eye detection like the older versions of the program. The changes thus made are reversible, and the original scan file is not altered when the effect is applied. Red-eye correction is one of the most important standard image processing tools, and its convenient inclusion in Capture NX 2 helps take another significant step towards life without Photoshop.

▶ A red-eye control point is sufficient to remove red-eyes from NEF RAW data.

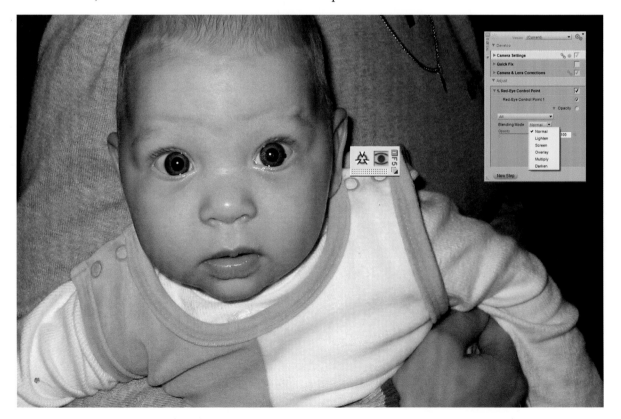

Color Boost

Color Boost is an easy tool to use and is available in Capture, but not in Nikon Scan. This tool's effect is the same as making a Chroma adjustment in the LCH tool, but is simpler to apply. The intensity of the colors in your image can be increased simply by moving the slider in the tool's dialog box, helping to make even the palest scan more vivid.

This tool is particularly effective when used with negatives. Slides are less prone to need such processing, especially if scanned using a high-quality device such as the Nikon Coolscan 5000. Using Color Boost can lead to the creation of color casts, and the tool's Protect Skin Tones option helps to counteract this effect, preserving natural skin tones even if other colors are (relatively) heavily adjusted.

Photo Effects

The Photo Effects palette includes tools for coloring RAW images in Black and White, Sepia, or Tinted tones. You can fine-tune your results using the Cyan–Red, Magenta–Green, and Yellow–Blue sliders.

▼ The browser of Capture NX2 reacts quite slowly; for pure viewing, you should use a specialized viewing program like Nikon View.

▲ With U Point Technology, dedicated
control points help you fine adjusting
color manipulation.

13.4 U Point Technology

Capture NX 2 does not support the conventional image layer system
that Photoshop uses. Instead it uses U Point technology licensed from
Nik Software. U Point-based tools are very powerful and much easier to
use than layer techniques. While the Nikon approach is simpler to use,
Adobe users still have more choice in the range of fine-tuning options.

Control Points Instead of Layers and Masks

U Point technology enables you to apply many basic changes to RAW
scan files that were previously only available for use with other data
formats. Here, the basic tool is the Color Control Point rather than the
Wand/Mask combination used in Photoshop. You can also choose from
Black, White, Neutral, and Selection Control Points when making your
adjustments.

Control Points recognize the position, color, saturation, and (according to the manufacturer) texture of the selected image area. If, for example, you wish to color a yellow part of your image red, all you have to
do is set a color Control Point in the yellow area and use the Color Picker
to display the current color and to select the desired new color.

◄ Selective color adjustments are easy with U Point technology, even for users with little experience. If you want only specific areas to be changed, you need several control points in most cases.

More Control Points = More Precision

It is not easy to apply color effects accurately using a single Control Point. The more Control Points you set, the more accurately you can delimit the area you apply your effect to. The manufacturer recommends using as many Control Points as possible. This may sound complicated, but our example demonstrates the opposite. For this example, I set 11 Control Points, and I was done with the recoloring within five minutes. Control Points really are easy to use, and especially useful for hobbyists and others who don't process images professionally.

No Masks, no Layers Needed

U Point technology makes complex layer and masking techniques obsolete. It also includes special Selection Control Points and the Show Selection option, allowing you to check and adjust your selection precisely for every Control Point you set.

Color Control Points can be set to run in various modes, including BCS (the default mode), RGB, HSB, and All. These modes control exactly which aspects of your image are affected by the Control Point you have set. BCS mode adjusts the brightness, contrast, and saturation parameters, while RGB mode simply adjusts the red, green, and blue color channels according to your settings. HSB mode adjusts hue, saturation, and brightness; and All mode does exactly what it says, adjusting all of the above-mentioned parameters during processing.

13.5 Grain, Noise, and Unsharp Masking

An Alternative to GEM

One Capture function that is aimed squarely at nostalgically-minded digital photographers is the Add Grain/Noise tool. Most people, however, wish to reduce grain while scanning, and Capture's improved Noise Reduction tool is a useful alternative to the GEM function found in most scan software. Similarly to the GEM plug-ins available for Photoshop (Noise Ninja, Neat Image etc.), you can use the tool to try out various effects without having to waste time rescanning.

The major advantage of Noise Reduction is that, like all Capture NX 2 tools, its effect on NEF files can be reversed at any time. The tool's effect on image detail can, however, be fairly brutal, and you may find that Noise Ninja allows you to preserve more detail while still effectively reducing noise.

New Filters

In addition to the well-known Unsharp Mask filter, Nikon has added High Pass and Gaussian Blur filters to the range of sharpness adjustments available in the program. The High Pass filter can be used to apply a quite aggressive sharpening algorithm to your images, which is particularly suitable for print applications. The Gaussian Blur tool has exactly the opposite effect; it applies significant softening to your scan. Similar filters are also available in Photoshop but are applied using the layer model, which is not part of the Capture NX concept. These filters are applied in NX by first configuring the filter itself and then setting the opacity with which the effect is applied to the actual image. Some people claim the Photoshop method is more straightforward, while others maintain that the world has simply had more time to get used to the Adobe way of doing things!

To conclude: Capture NX 2 is the best program currently available for quick and easy processing of NEF image archives. Photoshop, in spite of its slightly more cumbersome handling, still retains its lead when it comes to the power and range of its creative image processing tools.

NEF Files in NX 2: No Reverse Compatibility!

Capture writes all alterations directly to the NEF image file, and the NX and NX 2 versions of the software are unfortunately not reverse compatible. This means that once you have saved a NEF file in NX or NX 2, it can no longer be opened using the older Capture or Nikon Scan. Before saving a file using Capture NX or NX 2, please be sure that you will be processing the file exclusively with the same program in the future.

Correcting with Photoshop and Photoshop Plug-ins

Adobe Photoshop is the gold standard for professional image processing. The program has useful tools not only for creative image manipulation, but also for correcting typical flaws in scanned images. It allows you to straighten tilted horizons, perform perspective corrections, and remove red-eyes. There are a number of tools in Photoshop for scratch removal in black and white film: apart from SilverFast, no other scanner program has suitable tools for that. Plug-ins make Photoshop particularly useful for sophisticated special jobs, such as grain removal or color restoration of faded originals.

14

Contents

Uncorrected

Corrected

▲ The noise filter "Dust and Scratches" is a tool with noticeable side effects.

14.1 Dust and Scratch Removal

Under Filter ▶ Noise Photoshop offers Dust and Scratches. It affects the entire image, unless it is used with a selection or a mask. The only parameters available for the filters are Threshold and Radius. The preview shows how the weakest setting reduces the image sharpness. The effects of scratch removal are acceptable if the correction is limited to a radius of 1. Larger radii lead to unacceptable loss of quality. The filter is acceptable for fine scratches but is not suitable for deep scratches. From my own experience, for quality reasons, I can hardly recommend this method. The Clone Stamp or the Polaroid plug-in is better.

Removing Scratches with the Clone Stamp

A commonly used retouching tool in Photoshop is the Clone Stamp. To use this tool, first pick a suitable source region and copy the pixel pattern from that region to patch up the damaged areas. The size of the stamp is adjustable – it is possible to perform corrections down to a single pixel. In scanned images, scratches rarely occur in isolation; usually, many small scratches must be dealt with. Add to this the usual fine dust and fibers that are picked up by film, and you see there can be serious need for corrections, even with carefully handled film.

If you are a quality fanatic, you can remove all such defects using the Clone Stamp with excellent results. However, it is enormously time-consuming to do that. Because the Clone Stamp transfers 1:1, you must take great care in selecting your source area; otherwise, you will impair the texture and color of the target area. The Photoshop Healing Brush is

◀ Removing scratches with the Clone Stamp is very cumbersome and time-consuming.

◀ The Healing Brush is simpler to use than the Clone Stamp.

more flexible; but, when used properly, the Clone Stamp is hard to beat for quality. Even ICE costs you some sharpness, which is not the case with Clone Stamp. Cloning is much too cumbersome for regular dust and scratch removal after the scan. It is not practical for high-volume scanning.

Healing Brush and Spot Healing Brush

Besides the Clone Stamp, you can also use the Healing Brush for correcting dusty and scratched image regions. You still have to pick the source region yourself, but Photoshop calculates automatically what to use to fill the spot. Occasionally, you will get some unwanted results from the brush, such as having a color from a neighboring area pulled into the correction by mistake. In such cases, you can go back to the Clone Stamp. With the ALT key pressed, you first click on the area you want to use as the source for the corrections. Based on these data, Photoshop will calculate the correction. Then you can brush over the defective areas. The smaller the brush radius, the more precise the correction; but also, the more time-consuming the job for the whole image.

If this method were used to remove every single speck of dust – and the harsh LED lighting of the Nikon scanners captures a lot of them – retouching with the Healing Brush would be very laborious. However, fixing a small number of image defects with the Healing Brush is typically quick and easy.

> ## Since Photoshop CS2: Spot Healing Brush
>
> Both Healing Brush and Clone Stamp always require manual input of the source area. This time-consuming step is no longer needed with the new Spot Healing Brush: You simply activate the tool and immediately start to work. Spot Healing Brush is therefore an ideal retouching tool for big dust and scratch removal jobs.

Dust & Scratch Removal Plug-in from Polaroid

You can download a free plug-in for dust and scratch removal from Polaroid at www.polaroid.com/service/software/poladsr/poladsr.html. The plug-in is installed under Filter ▸ Polaroid ▸ Dust & Scratch Removal. Because it lacks before-after comparison and a preview for the entire image, it is not very user-friendly. However, you can get used to it; and it is, after all, a freeware tool. The job it does on dust removal for black and white films is also excellent. Even with only minimal configuration, the dust across the entire image will be eliminated. There is no need to mark the affected areas. The plug-in automatically detects areas of dust. With a weak setting, you can apply the filter to the whole image by activating the switch Light Dust. The scratch removal is not as powerful, but most software-based tools have a problem with scratch removal anyway.

Unfortunately, the position of the preview window of the Polaroid plug-in cannot be chosen freely. However, there is a little trick that can help matters. Simply select a particularly dusty area with the Rectangular Marquee – the selected part of the image will be displayed in the little preview window. If the selection is too large, the preview will only show the upper left corner of the area. When making a selection, note that the filter works for the selected area only.

Perfect for Fast Processing

This is especially useful when you try to get the hang of this plug-in, since you can see the difference between the corrected and uncorrected portions of the image. For dusty black and white film, this tool is a great help; it is much faster than the cloning method. But, for long scratches, some work will still need to be done with the Clone Stamp.

14.2 Image Corrections

Correcting Perspective Distortion

Perspective distortions happen when the camera is tilted upwards or downwards during shooting. For architectural photography, this is difficult to avoid, since the photographer usually stands on the ground and points the camera upwards. The distortion that is generated gives the impression that the building is tapering off toward the top. In most cases, you do not want this effect; therefore, you have to fix the image on the computer.

Nikon Capture does not allow perspective correction at RAW level, but in Photoshop it can be easily done with the Crop tool. For automatic correction, Photoshop needs the frontal view of a rectangular object that exhibits keystone distortion. The small inset in the following example shows a building with the walls becoming narrower toward the sky. Of course, in reality, walls and towers are rectangular.

First, draw a rectangular frame around the entire wall with the Crop tool. Then activate Perspective in the option bar. Now pull the corners of the rectangle to match the inclined walls. (To restrict the mouse movement to horizontal or vertical only, hold down the Shift key.) When that is done, Photoshop knows which area is supposed to be rectangular

▲ Polaroid's Dust and Scratch Removal plug-in is very well suited for automatically removing fine dust.

▲ The outer walls of the building are supposed to be straight (but in the insert they are obviously not). You can straighten them out with the Photoshop Crop tool.

and the actual shape of the area. Next, enlarge the crop to cover the entire image by pulling on the mid-points of the crop border. Finally, double-click in the crop area. This gives the corrected image a trapezoidal outline; simply crop again to get a rectangle.

Selectively Correcting Shadows and Highlights

Under Image ▸ Adjustments ▸ Shadow/Highlight, Photoshop offers a tool for selectively improving shadow and highlight areas. This feature is functionally identical to AACO from SilverFast and DEE/D-Lighting from Nikon. In the extended options, any of the three basic parameters (Shadows, Highlights, and Adjustments) can be tuned separately. Photoshop goes well beyond the capabilities of Nikon Capture in this regard.

The above image shows a backlit shot with blown-out highlights and pronounced shadow areas. The corrected version on the right clearly shows more detail. As usual in Photoshop, the correction is permanent; it cannot be undone at a later session. This can be a bit awkward if you are still trying to find the best setting for the correction. Nikon Capture is more user-friendly in this regard.

◀ On the left is the uncorrected image; on the right is the corrected image, after a combined Shadow/Highlight correction.

Straightening Horizon

With the tool Measure (under the Eyedropper pull-down menu) and the command Rotate Canvas, Photoshop offers a simple way to straighten inclined horizons. First, take the Measure tool and draw a line along the horizon. Next, go to Image ▸ Rotate Canvas and select the option Arbitrary. Photoshop takes the angle from the measuring line and enters it into the field for rotation. Click "OK", and the image rotates. Unlike the older Capture software, Capture NX can rotate an image to any desired angle. The Crop tool offers another way to rotate the image. First, select the whole image or a smaller crop. Then, left-click outside the crop frame and rotate it freely with the mouse.

14.3 Photoshop Plug-ins from Applied Science Fiction

The plug-ins introduced here come from Applied Science Fiction (ASF) and can be found at www.asf.com. The free trial versions are fully functional but put a watermark into the image. If some of the filters are grayed out, this indicates that the image was probably opened in an unsupported mode. Older versions of ASF plug-ins support only 8-bit; the newer versions also support 16-bit color depth, which makes them fully compatible with Photoshop CS, CS2, CS3 and CS4.

* An unwanted side effect of all the tools to correct Highlights/Shadows is an increase in image noise. These effects are best countered on a hardware level. Multi-Exposure scans significantly reduce image noise and have greater image data reserves for subsequent image processing.

▶ The Measure tool helps you determine the angle to rotate the horizon.

Grain Removal with Digital GEM

This plug-in is available in the versions Digital GEM, Digital GEM PRO, and Digital GEM Airbrush PRO. All versions deliver the basic functionality to remove film grain from an image. The plug-in Airbrush PRO belongs more to the artistic field. With the airbrush function, portraits can be modified to make people look like flawless creatures with porcelain skin.

Corrections using Nikon Scan's GEM are fixed into the image file. Any adjustments attempted later require a rescan. It is more convenient to work with the Photoshop plug-in, which allows you to experiment with different settings.

Digital Fill-Flash with SHO

Smoothing Grain While Using Scan Software

Nikon Scan is the only currently available scan software that includes an implementation of GEM. SilverFast has its own GANE (Grain and Noise Reduction) tool, and VueScan has a filter called Grain Reduction.

This plug-in comes with the versions Digital SHO and Digital SHO PRO. The PRO version is interesting because of its additional adjustment controls. Digital SHO corrects blown-out highlights and underexposed shadow areas by optimizing contrast and exposure. You can do the same thing with Curves, but not as easily. Digital SHO works just like DEE, D-Lighting, or Photoshop's Shadows/ Highlights.

Since Photoshop has already integrated the function Shadows/Highlights from version CS on, this SHO plug-in is worthwhile only for those who use older Photoshop versions. But in this case you should think about upgrading to a more current version of Photoshop to enhance your workflow.

◄ With the ROC plug-in, fresh colors can be extracted from faded film. The corrections can be set finer in the plug-in than in the scan programs.

Restoring Colors with Digital ROC

This plug-in is used to restore colors in scans from faded originals. It comes in two versions: Digital ROC and Digital ROC PRO. Only the PRO version supports 16-bit color depth. Also, Nikon Scan uses ROC (see chapter 7). VueScan and SilverFast offer comparable functions. Compared with other scanning programs, the Photoshop plug-in has the advantage because you can experiment with different settings without the need for a new scan each time. In addition, the plug-in has finer controls for color correction than the scan programs. This is the reason why you should choose the ROC plug-in if you can afford it. To work with the trial version is no option whatsoever, as it stamps all your pictures with big watermarks.

Plug-in vs. Scan Program

All the ASF plug-ins mentioned here are implemented in Nikon Scan; VueScan and SilverFast also have similar correction filters built in. Although they cost extra, the advantages of using plug-ins are obvious: You can try out the effects of various settings on your scan without having to continually make time-consuming new scans. Plus, in most cases, the plug-in has enhanced capabilities comparable to the scan programs. If you have the choice, you should take the plug-ins for these corrections and not the scan programs.

▶ Pure white does not have such different values for RGB; therefore this image has a color cast.

14.4 Removing Color Casts

Images that have the pure color tones black, white, or neutral gray are easy to correct. In each case, simply click the corresponding eyedropper for color correction. The real world is different, however, since you rarely shoot color charts in a laboratory. Therefore, you really need to consider applying Photoshop's color correction tools. To name them all would go beyond the scope of this chapter. In the appendix, there is a list of recommended Photoshop guidebooks that describe these techniques in detail.

Identifying Color Casts

Before a color cast can be corrected, it first must be identified. The Info tab (second tab in the Navigation toolbox) can be immensely helpful. In the photo, the cursor is on a white mussel. According to the RGB principle, the three color channels should have roughly identical values for pure white. This is not the case with the picture of the mussels; it obviously has a color cast.

Removing Color Casts

In most cases, you can improve a color cast by simply adjusting the levels of each individual color channel. This makes the entire image more contrasty and eliminates the color cast for the most part. Unfortunately, the adjustment will not work for all subjects.

This example shows a color cast caused by an uneven distribution of tonal values in the red and green channels. Broadening the tonal range for these two channels removes the color cast and increases general image contrast.

14.5 Match Color

Another useful tool in Photoshop CS2 for removing color casts is Match Color. Originally, this feature was developed to get consistent colors among the pictures in a series, but it also works fine for removing color casts. To do that, select the tool and click the checkbox Neutralize. In theory, the color cast is supposed to be removed automatically; but in most cases, the correction still requires some manual adjustments which are not too much trouble to perform. With the Fade slider, you can adjust the intensity of the correction to your needs. It can sometimes happen that the colors have been overly affected in color cast correction; with Fade, this can be fixed.

Overall, Match Color is a useful tool for selectively correcting color casts without having to deal with the different color channels.

▶ The uncorrected image of a coastline has a yellow cast.

▶ With the Fade slider, you can tune the strength of the correction.

14.6 **Noise Reduction with Noise Ninja**

Noise Ninja is a program for noise removal. It comes as a standalone program and as a Photoshop plug-in. Licenses cost between $35 and $80 and are available at www.picturecode.com.

The functionality of the tool is basically comparable to GEM, but Noise Ninja is gentler on the texture of the image. Especially with grainy slides, the erratic sprinkle of color dots often leads to unpleasant images after scanning. On the PictureCode website, there are many ready-to-use profiles for the grain patterns of popular scanners; but these are to be used with caution, since the profiles depend not only on the scanner model, but also on the combination of scanner and film.

The noise reduction often works very well, but its effectiveness depends very much on the image. Different sources on the Internet consider Noise Ninja to a better choice than GEM, but I do not fully agree with that evaluation. Each method has its own characteristics which you either like or don't like. As already mentioned, Noise Ninja is gentler on the texture of the image, but in all cases, corrections using noise reduction suffer a loss of detail as a side effect.

▼ The correction mostly removes the unwanted sleety grain texture.

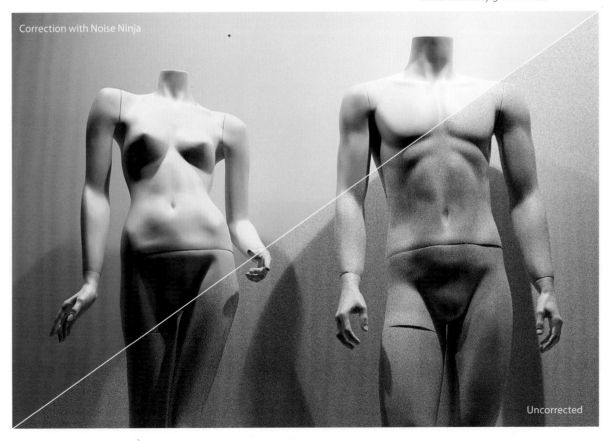

Correction with Noise Ninja

Uncorrected

14.7 Noise Reduction with Neat Image

Even More Noise Reduction
In addition to the tools mentioned here, there are many more noise reducing plug-ins available on the market, such as Nik Software's DFine, Akvis Noise Buster, or Imagenomic Noiseware.

Aside from GEM and Noise Ninja, another way to combat image noise is to use Neat Image. The Photoshop CS2 Reduce Noise feature offers an integrated function similar to Neat Image; but, in my tests, I found this filter rather useless.

Neat Image is available from www.neatimage.com for between $30 and $75. Its functions and results differ only slightly from those of Noise Ninja. Neat Image also allows you to apply scanner profiles, but the selection of scanner models on their website is somewhat meager. Much better are the profiles for cameras, for which this program has an excellent selection.

The resulting pictures from Neat Image and Noise Ninja are basically alike, since they both do less harm to the film's texture than GEM. In my tests on Nikon scanner files, I found that Noise Ninja usually gives better results than Neat Image. My general impression is that Neat Image reduces image sharpness more than Noise Ninja does. Then again, this effect depends heavily on the scanner model and the film material.

▼ Noise reduction always reduces image detail, an effect visible in the clouds in this example.

uncorrected

corrected with Neat Image

Backup

It sounds trivial to say that you need to backup your images. But in reality, this message does not sink in with most users–sometimes not even with computer-savvy users. Contrary to popular belief, certain storage media, such as DVD, are not at all suitable for storing valuable image data. This chapter examines common backup methods and media for their suitability as a digital image archive. If you have several thousand images, even with a well maintained filing system you could easily lose track of your image collection. IPTC data help bring order to the deluge of pictures.

15

Contents

15.1 Saved Does Not Automatically Mean Safe

Each file created on the computer is basically saved to the hard disk first. Now the file has been saved, but it is not safe. If a virus, a disk crash, a program error, or user error damages the file, it is gone. For scanned data, this may not be fatal, since the analog source is usually still available; but the time needed for rescanning can be considerable. In this case as well, backup is mandatory. Backup and management must be economical, simple, and effective. The more complicated the backup, the more likely it is to be neglected. Problems with backup can usually be traced to user error or infrequent use due to awkwardness.

15.2 Backup to Hard Disk

Any backup is better than no backup at all. On closer examination, the way backup is often done is usually less than ideal (e.g., backing up on a local drive). Backup may be convenient, but in most cases it is not really secure.

Backup to the Same Partition

Backing up files by duplicating them on the same hard drive partition as the original files gives no protection in the event of:
- Hard drive crash
- Computer theft
- Accidental formatting of the partition or the entire hard drive
- Virus attack on the local computer

Therefore, local duplication is merely a compromise for quickly saving an intermediate state. Backing up to a different partition on the same hard drive offers slightly more security, but it is still not a proper way to securely backup files.

Backup to an Additional Hard Disk

Somewhat better is backing up to a separate internal hard drive. However, this still does not protect files in the event of:
- Computer theft
- Virus attack on the local computer

The aforementioned applies to permanently installed internal hard drives, so they are still not ideal for backup. Removable hard drives that are connected to the computer only during backup are better. This type of backup also reduces the risk of theft, as long as the hard drives are stored in a different location from the PC. Also, the risk of a virus infecting a removable hard drive is lower than for an internal hard drive. This risk can be further reduced if appropriate access rights are set at the

Image: Western Digital

◄ If you want to transfer large amounts of data to an external hard disk, you need a fast interface. The older USB and FireWire standards are slowly being superseded by the more powerful eSATA technology.

operating system level. Additional hard drives can be mounted either with a swappable drawer system inside the PC or with connections as external drives. External disks can be connected with USB 2.0/Fire-Wire without measurable performance degradations. Unlike a drawer-mounted hard drive, an external drive can also be connected easily to other computers. Even using two or three external hard drives dedicated for backup is less expensive than an equivalent solution using backup tapes. A hard drive's mechanism is sensitive to shock; but with adequate care, this is not really a problem.

Forget about USB and FireWire

If you want to transfer large amounts of data quickly and safely, you are even better off using the newer eSATA technology rather than USB or FireWire. An external eSATA hard disk is nearly as fast as an internal hard disk and is an ideal storage platform for home users.

15.3 Archiving on Removable Media

There are many different types of storage media for all kinds of purposes. To make the backup process more manageable, ideally, the image archive should fit onto a single data carrier. With today's typical high-resolution scans, a total capacity of 100 to 200 GB for the entire image archive is rather low. Among other media, there are no optical discs available yet in this capacity range, and tape drive offerings are rather sparse.

Hard Drive as Removable Storage
External hard drives currently offer the lowest cost per GB of storage. They are as reliable as internal drives when treated with care. If necessary, several hundred GB can be backed up in reasonable time without having to swap media.

Optical Media: CD, DVD and Blu-ray

Optical media are cheap and easy to record, so they are ideal for removable storage. One problem is their relatively small storage capacity, at least for storing files from a film scanner. For example, a roll of 36 exposures scanned at 2900 dpi in 12-bit RAW format requires over 2 GB of storage space. A regular DVD with 4.7 GB capacity lets you store just two rolls of film at that resolution.

Regular tests in related magazines show that DVD recording technology has not yet fully matured. Current DVD burners show progress in burning speed; but with respect to data security, many burners and blank disks are not worth recommending. They are not suitable for reliably archiving your images. DVDs are more useful for data exchange, since they are inexpensive and almost all current computers can read them. They are also good for mailing.

There are no reliable findings on the long-term stability of recordable CDs or DVDs. For backup, they are neither safe enough nor do they have high enough capacity. The same is true of the newer Blu-ray technology. The medium's capacity of 50 GB per disk is significantly greater than that a DVD, but it is still not sufficient for storing complete archives of high-resolution scan files.

Image Tape

Because of their enormous storage capacities of 100, 200, 400, and 800 GB, LTO tapes are seemingly good candidates for maintaining large image archives. Unfortunately, the costs for such systems are enormous, too: Good drives start at around $1,200 and tapes are around $40 each, which is not exactly a bargain. Besides the cost of the drive, you have to invest in a fast computer, too.

LTO drives are designed for professional purposes, which means they are generally designed to run on a performance server. If you use a standard PC instead, this can lead to malfunction, as the drive will try to write the data faster than the system can deliver.

Images: Plextor, HP

Even Blu-ray's capacity of 50 GB per disc is not sufficient for storing complete image archives.

LTO Ultrium 3 streamer with 400 GB uncompressed capacity. From a technical standpoint, this is ideal for backing up image archives, but it is not inexpensive.

15.4 Image Management

For managing a large number of digital images, simple image browsers such as Nikon Capture or Photoshop are not sufficient. Programs such as ThumbsPlus or iMatch offer far more powerful cataloging functions.

Setting Up an Image Database – Folder and File Structures

The first step in setting up an image database in a meaningful way is to construct a logical folder structure. If you scan slides as well as negatives, the directory tree on your hard disk should factor in the structure of your analog archive. This will help you to associate a file easily with the analog original. For a rescan, you will not have to search very long to find the correct slide or filmstrip. The following information can help simplify retrieval:

> *A well-planned folder structure is usually sufficient if you want to index a small photo archive. This way you can save yourself the effort of individually annotating your images with IPTC keywords.*

- ▸ Year of shooting (e.g., 2005)
- ▸ Month and day of shooting (e.g., 08-27)
- ▸ Unique naming of the film with sequential numbering (e.g., F001)
- ▸ Film strip number or position in the slide tray (e.g., 36A)
- ▸ Occasion (e.g., Wedding André/Dani)

This type of information should be reflected in the folder tree of the hard disk, the IPTC information of the images, and the filing system of the physical film archive. A well-structured filing system speeds up your search. In addition to the logical directory structure, you can include key information such as the date in the file name. If you move or copy files, this type of naming convention makes it easy to reconstruct the origin of those files. With this method, you will notice errors immediately when moving files. A file name such as 2004-10-Film01-Cliffs-Neg12.nef is much more descriptive than IMG12.nef (as generated by default in Nikon Scan).

The Simpler, the Better

There are no hard and fast rules for organizing your photo archives. A good yardstick is how long it takes a stranger to understand your system. The easier your system is to understand, the better it is.

◂ A chronologically structured file system simplifies retrieval in Windows Explorer.

▲ IPTC metadata is a useful functionality for cataloging even big archives.

Cataloging with IPTC Data

Some labels can be applied to an image with appropriate file naming conventions based on file and folder structure. However, quite often there is a need for additional information about images, such as:

▸ Who the photographer is; who owns the image rights
▸ Where and when the picture was taken
▸ What film type was used; what camera equipment was used

This list of metadata – data about data – could be extended infinitely. Metadata describe the image in as detailed a fashion as possible. For this purpose, text information is added to the image information and stored in the image file. The International Press and Telecommunication Council has adopted the IPTCNAA standard, or IPTC for short. Not all image file formats support this metadata entry. Suitable image databases, such as iMatch, offer quick searching of large numbers of files for a particular image.

The quality of the search depends on the quality of the IPTC metadata. A keyword search works only if each image file has the corresponding entry in its IPTC header. Many viewers and most image editors support IPTC for the standard formats JPEG and TIFF. The support for NEF format, however, is not as good. IPTC data can only be edited by programs that have full access to NEF, such as Capture NX2.

Batch Editing IPTC Information
To manage an archive, it can be tedious to enter IPTC image data for each image one-by-one. In an image sequence, much of the data can be repeatedly used for several images. With Nikon Capture, information for multiple images can be filled out at once. This function is somewhat hidden under Batch/Copy IPTC Info. It allows the IPTC entries of one image to be copied and subsequently applied to the target images with Batch/Paste IPTC Info.

Organizing Images with ThumbsPlus

In order to manage a large number of images efficiently, you need a good viewer with a thumbnail indexing function and its own image database, such as ThumbsPlus. Common viewers lack these functions. Nikon View, for example, generates the thumbnails anew each time it opens a folder. If you frequently jump between folders, the long wait required to generate thumbnails will test your patience. Especially for large scan files, the wait is considerable. ThumbsPlus solves this problem by generating the thumbnails only once: when the folder is opened the first time.

Performance Gain Through Caching

To do that, ThumbsPlus searches for all image files on the hard disk, generates thumbnails, and stores them in a separate database. This process takes some time, but it runs in the background while other work continues on the PC. Once the thumbnail index is complete, you can open a folder and immediately see the thumbnails. If you delete a file with ThumbsPlus, it updates the thumbnail database accordingly. If you delete images with a different program, such as Windows Explorer, ThumbsPlus needs to regenerate the index to keep the database up-to-date.

▲ The first time ThumbsPlus opens a folder, it generates thumbnails of the images inside.

RAW Support in Viewer Applications

Only a few viewers support RAW in their basic version. To display NEF files in ThumbsPlus, you need the Digicam RAW Plug-In. This plug-in for ThumbsPlus is noticeably slower than Nikon View NX for full-screen display of NEF files. If you use ThumbsPlus for managing NEF files, but use Nikon View NX for viewing individual images, you can combine the advantages of both programs quite nicely.

Batch Renaming Files: LupasRename

If you reorganize your folder and file structure, you may have to rename multiple files. With Windows Explorer this can be tedious, since you have to do it one by one. LupasRename is a lean tool for simple batch renaming. This freeware is available at http://rename.lupasfreeware.org. It helps if the files are already numbered sequentially. The names should contain a number that corresponds to the analog original. For example, Negative #21 would best be saved as IMG21.tif, so that you can keep the default names during scanning. Once the final name is found, you can

▼ Renaming groups of files in one go can be done quickly with LupasRename.

rename it with LupasRename. This tool supports all data formats, since it does not write into the file itself.

Glossary

AACO

"Auto Adaptive Contrast Optimization" from Silver-Fast. See also DEE.

Black Point

The darkest pixel in the image. In the levels curve, it is at the left end.

Brightness

A measure of the amount of reflected or directly emitted light. See also Luminance.

CMYK

Subtractive color model. Mixing the primary colors cyan (C), magenta (M), and yellow (Y) theoretically results in a deep black. However, in practice, this rarely works because the primary colors used are not absolutely pure. For that reason, the color black is added (the "K" stands for "key", which is short for "key plate," a term from the printing press). Printers work in CMYK mode.

Colorimeter

Device to measure color characteristics. Used, for example, to measure the color reproduction of monitors for the purpose of generating individual ICC profiles.

Color Management

Color management ensures consistent color representation across different devices. The monitor is supposed to display exactly the colors read by the scanner, and the print is supposed to look exactly like the monitor display.

Color Saturation

A highly saturated image appears colorful. Low saturation makes the image look flat due to the high content of gray.

Color Space

See Gamut.

Color Temperature

Reference to the temperature (in Kelvin) of a heated black body radiator, the hue of which is compared to the light of an incandescent lamp. High Kelvin values appear cold, and low Kelvin values appear warm to the human eye.

ColorSynch

The color management system of Apple Macintosh computers.

DEE

"Digital Exposure Extender" – tool for selectively correcting highlight and shadow areas. Suitable for recovering overblown highlights and brightening shadows.

Device-Dependent Color Space

Color space for a specific device (e.g., Scanner-RGB in Nikon Scan).

D-Lighting

Similar to DEE (see above), but applied to NEF RAW files afterwards. D-Lighting is fully reversible.

DPI (or dpi)

Dots per inch. Meant for printing, but also commonly used for monitors and scanners.

FARE

"Film Automatic Retouching and Enhancement" – hardware-based dust and scratch removal system of Canon scanners.

Gamut

Color range of an image or a color space. The gamut is the entirety of visible colors that a device (such as a monitor, printer, scanner, or film) can reproduce.

GANE

"Grain And Noise Elimination" system from Silver-Fast. See also GEM.

GEM

"Grain Equalization and Management" from Applied Science Fiction. Also used by Nikon Scan.

Grain Reduction

Grain reduction system from VueScan.

ICC Profile

The basis for color management systems, ICC profiles define the device-independent representation of colors.

ICE

"Image Correction and Enhancement" – hardware-based dust and scratch removal system from Applied Science Fiction. Also used by Nikon Scan. ICE relies on the infrared light source of the scanner.

ICM

"Image Color Matching" – the integral color management module of current Microsoft operating systems.

Infrared Clean

Hardware-based scratch removal system from VueScan.

iSRD

Similar to SRD (see below). Unlike SRD, iSRD uses the infrared light source of the scanner to detect dust and scratches.

IT8 Target

A standard test target for profiling devices, such as scanners. SilverFast ships with an IT8 target in the form of a slide.

LCH

"Lightness, Chroma, Hue" – a color model describing colors in terms of those three optical parameters.

Luminance

Brightness or intensity of a color (in the context of this book).

PPI (or ppi)

"Pixels per inch" – used for images displayed on monitors.

Printer Profile

Profile describing the color response of a particular printer with a specific paper and a specific ink. It is used to produce prints with accurate colors.

Restore Fading

VueScan's equivalent of ROC. See ROC.

RGB

"Red, Green, and Blue" – in the (additive) RGB color model, all colors are generated from red, green, and blue.

ROC

"Restoration of Colors" – a system from Applied Science Fiction for restoring the colors of faded film. Also used by Nikon Scan.

Scanner Profile

An ICC profile describing the color characteristics of a particular scanner to allow a correct interpretation of the colors of the scanned image.

SRD

"Smart Removal of Defects" – a software-based system for dust and scratch removal from SilverFast.

Working Color Space

Soft Color space describing the colors of the image independent of the device used (e.g., film scanner, monitor, and printer). The working color space used for photos is usually Adobe RGB (1998) or sRGB.

White Point

The brightest level in an image (not necessarily the color white). When setting the white point to a particular input level, this level will be corrected to become the highest possible output level. All higher input levels will be lost, since the corresponding pixels change to the color of the white point.

Resources

[1] **filmscanner.info**
On Patrick Wagner's home page, you can find
test reports for most current film scanners
from his online shop. This is great for getting
an overview of the current scanner market. In
English, German, French, Spanish, and Italian.
www.filmscanner.info

[2] **medienarchiv.com**
Reviews of film scanners and software. Contains helpful tutorials about scanning whole
archives. Unfortunately only for German
speakers.
www.medienarchiv.com

[3] **Depth of Field Calculator**
This web page from Erik Krause contains several interesting tutorials for image processing,
as well as a depth of field calculator. The site is
for German and English speakers.
www.erik-krause.de

[4] **scantips.com**
Website about various scanning techniques,
with tutorials. This is great for beginners, but
advanced users will not learn much.
www.scantips.com

[5] **SilverFast – The Official Guide**
In this book, Taz Tally summarizes information that you can already get from SilverFast
PDF documentation. Print quality and images
are not up-to-date.
www.sybex.com

[6] **The Photoshop CS3 Book for Digital Photographers**
This book by Scott Kelby guides you step-
by-step through many different digital photography techniques without burdening you
with any theory. It's ideal for teaching yourself
Photoshop (learning by doing). Kelby doesn't
go deep into the basics, but this is a good book
for using Photoshop to turn your scan into
pretty pictures.
shop.scottkelbybooks.com

[7] **Adobe Photoshop CS4 for Photographers**
This is heavy stuff (700 pages, $45), but Martin
Evening provides a very deep and thorough
introduction to Photoshop. It is not as easy to
read as Scott Kelby's book, but it teaches you
not only what you have to do but also why you
do it.
photoshopforphotographers.com

[8] **slidescan.info**
On this website you will find additional
information about this book, provided by the
author Sascha Steinhoff. Here he answers questions from readers and offers a huge download
section with high resolution sample scans.
www.slidescan.info

DVD

Scan Software
VueScan (Linux/Windows/Mac)

Picture Editing
Adobe Photoshop CS4 (Windows)
SilverFast HDR (Windows/Mac)
Photoshop Plug-ins (Windows/
Mac)

Image Databases/Viewer
Irfanview (Windows)
ThumbsPlus (Windows)
IMatch (Windows)

Sample Scans Filmscanner
Braun Multimag SlideScan 4000
Microtek Artixscan 120TF
Nikon Coolscan 5
Nikon Coolscan 5000
Plustek OpticFilm 7200i
Quato IntelliScan 5000
Reflecta ProScan 4000

Sample Scans Flatbed Scanner
Canon 9950F
Epson Perfection V750 Pro
Epson 4990
Epson F3200
HP Scanjet 8200
Microtek i900
Plustek OpticProST64+

Index

Scanning Correction Filters in Scan Programs and Photoshop

Name	Software	Before Scanning	After Scanning
Dust and Scratch Removal			
ICE	Nikon Scan, SilverFast and other	yes	no
Infrared Clean	VueScan	yes	yes (RAW)
FARE	Canon Filmget	yes	no
iSRD	SilverFast	yes	no
SRD	SilverFast	yes	yes (HDR)
Clone Stamp	SilverFast	yes	yes (HDR)
Clone Stamp	Photoshop	no	yes
Healing Brush	since Photoshop CS	no	yes
Spot Healing Brush	since Photoshop CS2	no	yes
Polaroid Dust & Scratch-Plug-in	Photoshop	no	yes
Color Restauration			
ROC	Nikon Scan	yes	no
ROC Plug-in	Photoshop	no	yes
Restore fading	VueScan	yes	yes (RAW)
Grain Removal			
GEM	Nikon Scan	yes	no
Neat Image	Photoshop-Plugin	no	yes
Noise Ninja	Photoshop-Plugin	no	yes
GEM Plug-in	Photoshop	no	yes
Grain reduction	VueScan	yes	yes (RAW)
GANE	SilverFast	yes	yes (HDR)
Shadows/Highlights			
DEE	Nikon Scan	yes	no
D-Lighting	Nikon Capture	no	yes
AACO	SilverFast	yes	yes (HDR)
Shadows/Highlights	since Photoshop CS	no	yes
SHO Plug-in	Photoshop	no	yes
Red Eye Removal and other Tools			
Red Eyes	Nikon Capture	no	yes (even with NEF)
Red Eye Tool	since Photoshop CS2	no	yes
Match Color	since Photoshop CS	no	yes
Straighten	Nikon Capture	no	yes (even with NEF)